A Symmetry Primer
for Scientists

A Symmetry Primer
for Scientists

||||||||||||||||||||||||||||||||||||||||||||||||

JOE ROSEN

Tel-Aviv University

A Wiley-Interscience Publication

JOHN WILEY & SONS

New York   Chichester   Brisbane   Toronto   Singapore

Copyright © 1983 by John Wiley & Sons, Inc.

All rights reserved. Published simultaneously in Canada.

Reproduction or translation of any part of this work
beyond that permitted by Section 107 or 108 of the
1976 United States Copyright Act without the permission
of the copyright owner is unlawful. Requests for
permission or further information should be addressed to
the Permissions Department, John Wiley & Sons, Inc.

Library of Congress Cataloging in Publication Data:

Rosen, Joe.
    A symmetry primer for scientists.

    "A Wiley-Interscience publication."
    Bibliography: p.
    Includes index.
    1. Symmetry. 2. Groups, Theory of.   I. Title.
Q172.5.S95R674 1982   500   82-10876
ISBN 0-471-87672-0

Printed in the United States of America

10 9 8 7 6 5 4 3 2 1

For Dalia and Laliv

Preface

||||||||||||||||||||||||||||||||

This book is intended to fill a gaping gap in the symmetry literature: Until its appearance there was, to the best of my knowledge, not a single textbook devoted to the fundamentals of symmetry and its application in science, in glaring contradiction to the undisputed importance of symmetry considerations in modern science. It is true that wherever its importance is greatly appreciated there exist textbooks on the application of symmetry in those fields, especially in quantum mechanics, crystallography, and chemistry. But the fact that these textbooks are almost always couched in terms of group theory and its application, rather than symmetry and its application, is symptomatic of the underlying problem that this book is intended to alleviate: Just as technical mastery of the Schrödinger equation does not imply commensurate understanding of the physics represented by that equation, neither does proficiency in the application of group theory in some field of science imply insight into the symmetry considerations of which that group theory is the formal expression. In fact, all too many science students, science teachers, and even scientists are uncomfortable, to say the least, with symmetry considerations. Students tend to be suspicious of symmetry arguments. They all too rarely take advantage of them and then all too often do it wrong. And when the lecturer, textbook, or thesis advisor uses a symmetry argument in a proof or to cut swiftly through to the solution of a problem, students feel cheated; something was pulled out of a hat. Science teachers often do not have a clear understanding of symmetry considerations, so they tend to avoid them, except when a symmetry argument is very obvious and compelling. And even those teachers who do understand and appreciate symmetry considerations rarely convey to their students a feeling for, if not some understanding of, the principles of symmetry and their generality and importance. Then we have the scientists, too many of whom, because they are uncertain about how to use symmetry considerations correctly, tend to be doubtful of them, especially of their own. And these doubtful scientists might at the same time be very proficient at the group-theoretical techniques involved in the application of symmetry in their own fields.

So what is sorely needed is a basic textbook on symmetry and its application in science, starting from scratch and presenting the material in an orderly and coherent manner, with problems for the reader to solve and a bibliography. Such a book should be able to serve as the primary text for a course on symmetry, as a supplementary text for science courses in which symmetry considerations play an important role, and as a self-study text for scientists, science teachers, and advanced science students who want to fill in what they may have missed and increase their symmetry sophistication. Detailed specific applications need not be included, since they can be found in abundance elsewhere. As you might suspect, this book attempts to fill just that role.

As for starting from scratch, the formal presentation of symmetry, of group theory, the mathematical language of symmetry, and of the principles of application of symmetry does indeed do so. However, for some of the additional material and for examples and problems throughout the book the reader is assumed to possess a certain background familiarity with various mathematical subjects and with various physical phenomena. For example, some familiarity with elementary algebra, complex numbers, geometric concepts, linear algebra, or ordinary differential equations is assumed at various points. And, especially, for the material on quantum systems a more than introductory understanding of quantum theory is required.

Thus, in keeping with its many roles, not all of the book is accessible to all readers. The undergraduate science student should be able to handle all of the core material except the quantum material and some of the examples and problems. The advanced graduate student should find almost all of the material, examples, and problems accessible, as should the science teacher and scientist. And any reader, upon returning to this book after gaining experience and insight in science, should find that more of it has become accessible and some of it has become clearer.

The book is structured as follows. Chapters 1 and 2 are an introduction to group theory. They present a reasonable dose of group theory and certain other mathematical concepts and supply the reader with the mathematical concepts and language necessary for the succeeding chapters, where symmetry is discussed in the language most suitable for it, the language of groups. For Chapters 1 and 2 to serve as a useful introduction to group theory they go somewhat further than is strictly necessary for the purpose of this book. Chapter 3 presents the general theory of symmetry, applicable to all, not necessarily even physical, systems. A special section on symmetry of quantum systems is included. Chapter 4 teaches the theory of application of symmetry in science. It contains some discussion of a philosophical nature, but most of it is utilitarian, and numerous examples are included. These first four chapters make up the core material of the book. Chapters 5 and 6, which require a more sophisticated reader than does most of the core material, Chapter 6 more than Chapter 5, are somewhat "extracurricular." They are not necessarily less important, however, and the first four sections of Chapter 6, discussing symmetry of the laws of nature and of initial and final conditions, symmetry in processes, and conservation, definitely should be read by the sufficiently sophisticated reader.

Numerous problems are included in this book, following almost every section of the first four and the sixth chapters. The reader should, of course, attempt to solve them in order to enhance the learning process and for self-testing. The problems of Chapters 1 and 2, a few of which extend the presentation, should all be accessible to whomever the chapters are accessible to. The problems of Chapters 3, 4, and 6, on the other hand, are not all accessible to every reader. Some of these problems purposely leave leeway for the reader's interpretation and thus do not have unique solutions.

A rather extensive bibliography is included, whose principal purpose is to offer sources for parallel, supplementary, complementary, and subsequent reading. Being very self-contained, this book rarely makes direct reference to the bibliography. When it does, a reference is designated by square brackets, [. . .].

For the person desiring preparatory reading to this book the cupboard is almost bare. As a nontextbook introduction to symmetry I dare suggest my own *Symmetry Discovered: Concepts and Applications in Nature and Science* [S17]. It is the only one I know of at its level and, like this book, was also intended to fill a gap in the symmetry literature. For a more advanced nontextbook introduction to symmetry I highly recommend Hermann Weyl's modern classic, *Symmetry* [S24].

A symmetry course based on this book should be structured as follows. Start with Chapters 1 (Group Theory, The Mathematical Language of Symmetry) and 2 (Group Theory Continued), if the students need an introduction to group theory. For a thorough introduction use all of both chapters. For a brief introduction that is sufficient for the rest of the course the essential sections are in Chapter 1: 1.1 (The Group Concept), 1.2 (Mapping), 1.3 (Isomorphism), 1.4 (Equivalence Relation), 1.6

(Subgroup), and in Chapter 2: 2.5 (Generators), 2.7 (Permutations, Symmetric Groups), and the material on equivalence class in Section 2.1. If the students have already been sufficiently introduced to group theory, mapping, and equivalence relation, these chapters may be skipped, unless a rapid review is desired. In no case should one attempt to teach the subsequent chapters to students who are not sufficiently familiar with group theory, mapping, and equivalence relation. Then present Chapters 3 (Symmetry) and 4 (Application of Symmetry) in that order. The last section of each chapter is only for students with a good formal background in quantum theory. After that Chapters 5 (Approximate Symmetry and Spontaneous Symmetry Breaking) and 6 (Symmetry in Processes, Conservation, and Cosmic Speculations) are for your picking and choosing. For these chapters a rather advanced scientific background is expected of the students.

The most important feature of this book, it seems to me, is the statement and rigorous derivation of six principles of symmetry: the equivalence principle (Section 4.2), the symmetry principle (Section 4.3), the equivalence principle for processes in isolated physical systems (Section 6.2), the symmetry principle for processes in isolated physical systems (Section 6.2), the general symmetry evolution principle (Section 6.2), and the special symmetry evolution principle (Section 6.3). I know that the equivalence principle, the symmetry principle, and the special symmetry evolution principle have been stated previously in one version or another [A17, A5, M36]. About the other three principles I do not know. In any case I am sure that this is the first time all the principles have been presented together, in their full generality, and within a coherent framework. And I am sure also that this is the first time the principles have been rigorously derived from a premise as fundamental as the very existence of science. The following diagram shows the logical structure of the derivations.

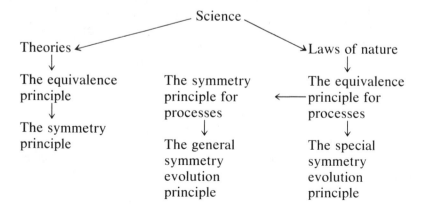

I want to express my deep thanks to the members of the Department of Physics and Astronomy of the University of North Carolina at Chapel Hill, especially Henk Van Dam and Jim Crawford, and to Kate Stern for doing so much to make my 1976–1977 sabbatical year, during which much of this book was written, so enjoyable and fruitful. I am grateful to Eugen Merzbacher for kindly reading and commenting on a preliminary version of this book and to Shimona Ginzburg for useful discussions on symmetry.

JOE ROSEN

Tel-Aviv
June 1982

Contents

||||||||||||||||||||||||||||||||||||

A Symmetry Primer
for Scientists

CHAPTER ONE

Group Theory, The Mathematical Language of Symmetry

In this chapter and the next, group theory and other mathematical concepts that are important for a serious study of symmetry are discussed. If you feel you are already familiar with the material, you might use these chapters for a review. Although it is desirable to master both chapters before continuing on to the next chapters, that is not absolutely essential. If you want to abbreviate your introduction to group theory, the sections that must not be missed are, in chapter 1: 1.1, The Group Concept; 1.2, Mapping; 1.3, Isomorphism; 1.4, Equivalence Relation; 1.6, Subgroup; and in chapter 2: 2.5, Generators; 2.7, Permutations, Symmetric Groups; and the material on equivalence class in Section 2.1. But if you want to understand group theory, you should definitely study all of both chapters.

This and the next chapter are only an introduction to group theory. There is very much more, and there are many books to help the interested student in his or her quest for further understanding. The bibliography at the end of this book lists a considerable number of books for more advanced study. As you might suspect, a more advanced study of symmetry than is presented in this book requires a more advanced study of group theory than these two chapters offer. The bibliography lists many books that take up the study of symmetry more or less where this book leaves off, and many of them contain correspondingly more advanced presentations of group theory.

If we were studying linear vector spaces, we could start by reviewing the familiar properties of ordinary vectors (2-vectors and 3-vectors), and, except for possible complications due to complex numbers and infinite dimensionality, we could use ordinary vectors as an excellent reference model throughout our presentation to help us keep our feet on the ground. But what can we do if we are studying group theory? How can we keep our feet on the ground? My own experience has led me to believe that the best solution is to leave the ground and start flying. So we will start with a completely abstract approach to groups and, with the help of examples and problems, try to develop a firm understanding of what is going on. (After all, ordinary vectors are also abstractions. It is only a matter of familiarity.)

This and the next chapter are self-contained and start very much from scratch. For the theory it would be useful to know a little set theory, but that is not essential. For the examples and problems some elementary mathematical knowledge is assumed, including concepts from algebra, functions, geometry, analytic geometry, complex numbers, and linear algebra.

1.1. THE GROUP CONCEPT

A *group* is, first of all, a *set* of *elements*. In a rigorous presentation one first studies some set theory before approaching group theory. (You might find it interesting to read some of the chapter on set theory in a book on abstract algebra.) However, for our purposes it is sufficient to know that a set is a collection and the elements of a set are whatever comprise the collection. An element of a set is said to be a member of the set, or to belong to the set. Besides whatever property the elements of a set have that makes them belong to the set, they need have no additional properties. For a set to be a group, however, very definite additional properties are needed, as we will see in the following discussion.

The number of elements of a set or group is called its *order*. It may be finite, denumerably infinite, or nondenumerably infinite. You know what finite means, and you most likely have a reasonable idea of what infinite means. *Denumerably* infinite means that the elements can be labeled by the natural numbers, 1, 2, 3, ... So the set of natural numbers is by definition denumerably infinite. The set of all integers (positive, negative, and zero) is also denumerably infinite, and so is (believe it or not) the set of all rational numbers (numbers expressible as the ratio of two integers). *Nondenumerably* infinite means that the elements cannot be labeled by the natural numbers. The set of all real numbers, for example, is of nondenumerably infinite order. (You might want to read about infinities, comparing infinities, greater and lesser infinities.)

We use capital letters to denote sets and groups and small letters for elements, in general. Curly brackets, { }, denote the set of all elements indicated or defined within them. Thus the equation

$$S = \{a,b,c,d\}$$

means that the set denoted by S consists of the elements denoted by a, b, c, d. The sign = in such equations means that the set on the left-hand side and that on the right are one and the same. Another example is

$$P = \{\text{all real numbers } a \text{ such that } -1 < a < +1\},$$

which means that the set P consists of all real numbers greater than -1 but less than $+1$.

The symbol $\in$ is used to indicate membership in a set or group. The statement

$$w \in U$$

means that the element w is a member of the set U.

Another use of the sign = is to relate elements of a set, as in the equation

$$a = b.$$

Here the = sign means that the two elements are one and the same element.

For a set to be a group it must be endowed with a *law of composition* (also called *law of "multiplication"*; note the quotes; this is not ordinary multiplication, so "composition" is preferable to "multiplication" to avoid confusion). This means that any two elements of the set can be combined (without specifying just how this is to be done). In fact, any pair of elements can be combined in two ways. If a, b are elements of G, the two compositions (or "products") of a and b according to the law of composition of set G are denoted ab and ba. (It is fortunate in terms of notation that composition involves only two ways of combining!)

But in itself composition does not yet a group make. To be a group, a set G together with its law of composition must satisfy the following four conditions:

1. *Closure.* For all a, b such that $a,b \in G$, we have

$$ab, ba \in G.$$

This means that for all pairs of elements of G both compositions are themselves elements of G. Another way of stating this is that the set G is closed under composition.

2. *Associativity.* For all a, b, c such that $a,b,c \in G$, we have

$$a(bc) = (ab)c.$$

This means that in the composition of any three elements the order of combining pairs is immaterial. Thus one can evaluate abc by first making the composition $bc = d$ and then forming ad, corresponding to $a(bc)$, or one can start with $ab = f$ and then make the composition fc, corresponding to $(ab)c$. The results must be the same element of G, which can thus be unambiguously denoted abc. In short, one says that the composition is associative. It then follows that associativity holds for composition of any number of elements.

3. *Existence of Identity.* G contains an element e, called an *identity* element, such that

$$ae = ea = a$$

for every element a of G. The characteristic property of an identity element is, therefore, that its composition in either way with any element of G is just that element itself.

4. *Existence of Inverses.* For every element a of G there is an element of G, denoted a^{-1} and called an *inverse* of a, such that

$$aa^{-1} = a^{-1}a = e.$$

In other words, for every element of G there is an element whose composition with it in either way is an identity element. (It might happen that $a^{-1} = a$ for some or all elements of a group.)

That ends the definition of a group. To summarize, a group is a set endowed with a law of composition such that the conditions of (1) closure, (2) associativity, (3) existence of identity, and (4) existence of inverses hold. A group is *abstract* if its elements are abstract, i.e., if we do not define them in any concrete way.

In general $ab \neq ba$ in a group. (Otherwise why insist on the two ways of composition?) But it might happen that certain pairs of elements a, b of G do obey $ab = ba$. Such a pair of elements is said to *commute*. From the definition of a group it is seen that an identity element e commutes with all elements of a group and that every element commutes with its inverse. Obviously every element of a group commutes with itself. If all the elements of a group commute with each other, i.e., if $ab = ba$ for all elements a, b of G, the group G is called *commutative* or *Abelian* (after the Norwegian mathematician Niels Henrik Abel, 1802–1829—but don't conclude that everyone who studies group theory dies so young).

In condition 3 we demand the existence of an identity element but do not demand that it be unique. And in condition 4 and the preceding paragraph we carefully refrain from referring to *the* identity in order not to imply uniqueness. However, all that deviousness was only to allow us the pleasure of *proving* that the identity is unique, as it in fact is. This will be our first example of a group-theoretical proof. Please note, and follow our example in your own group-theoretical proofs, that we will be very careful to justify each operation and each equation by reference to the definition of a group. This is especially important, since we are using familiar notation (multiplication, parentheses, equality) but assigning it new significance, and one might have a tendency to revert to old ways.

To prove uniqueness of the identity, we assume the opposite, that more than one identity element exist, and show that this leads to a contradiction. Denote any two of the assumed identity elements by e' and e''. By condition 3 we then have

$$e'a = ae' = e''a = ae'' = a$$

for every element a of G. Evaluate the relation $e'a = a$ for the specific element $a = e''$. We obtain $e'e'' = e''$. Now evaluate $ae'' = a$ for $a = e'$. We get $e'e'' = e'$. Comparing these two results, we have $e' = e''$, which is in contradiction to our assumption that e' and e'' are different. This proves the uniqueness of the identity.

Although we do not require it in condition 4, the inverse of an element is unique; i.e., for every element a of G there is only one element, denoted a^{-1}, such that

$$aa^{-1} = a^{-1}a = e.$$

As an additional example of group-theoretical proof we prove this statement. We again assume the opposite and show that it leads to a contradiction. So assume that element a has more than one inverse and denote any two of them by b, c. Then by condition 4

$$ab = ba = ac = ca = e.$$

Now according to the associativity condition

$$c(ab) = (ca)b.$$

By the equation before last the left-hand side of the last equation is

$$c(ab) = ce = c,$$

while the right-hand side is

$$(ca)b = eb = b.$$

(We use the property of the identity here.) Therefore $b = c$, and we have a contradiction to our assumption that b and c are different.

From the definition of the inverse and its uniqueness it is clear that the inverse of the inverse is the original element itself, i.e., the inverse of a^{-1} is a. This is written

$$(a^{-1})^{-1} = a.$$

The inverse of a composition of elements is the composition in opposite order of the inverses of the individual elements. In symbols this means that

$$(ab)^{-1} = b^{-1}a^{-1},$$
$$(abc)^{-1} = c^{-1}b^{-1}a^{-1},$$
$$\text{etc.}$$

This is verified directly. For example, we show that $b^{-1}a^{-1}$ is indeed the inverse of ab by forming their composition and proving that it is the identity

$$
\begin{aligned}
(ab)(b^{-1}a^{-1}) &= a(bb^{-1})a^{-1} && \text{(by associativity)} \\
&= aea^{-1} && \text{(by inverse)} \\
&= (ae)a^{-1} && \text{(by associativity)} \\
&= aa^{-1} && \text{(by identity)} \\
&= e && \text{(by inverse)}.
\end{aligned}
$$

The verification of

$$(b^{-1}a^{-1})(ab) = e$$

is done similarly.

We define powers of elements by

$$a^2 = aa,$$
$$a^3 = a^2a = aa^2 = aaa,$$
$$\text{etc.,}$$

and

$$a^{-2} = a^{-1}a^{-1} = (a^2)^{-1},$$
$$\text{etc.}$$

Then the usual rules for exponents are largely applicable, except that noncommutativity must be kept in mind. For example,

$$(ab)^2 = (ab)(ab) = abab \neq a^2b^2.$$

The *structure* of a group is the statement of the results of all possible compositions of pairs of elements. For finite-order groups this is most clearly done by setting up a *group table,* similar to an ordinary multiplication table (Fig. 1.1). To find ab we look up a in the left column and b in the top row; the composition ab is then found at the intersection of the row starting with a and the column headed by b. For the composition ba it is b which is on the left and a at the top. When the symbols for the group elements have the same ordering in the top row (left to right) as in the left column (up to down), a group table will be symmetric under reflection through the diagonal if and only if the group is Abelian. This is easily seen from the figure. Note that by reordering the rows or columns or by changing the symbols denoting the group elements a group table can be made to look different while still describing the same group. This must be taken into account when comparing group tables. Thus two group tables describe the same group, i.e., express the same structure, if they can be made identical by reordering rows and columns and redenoting elements.

Figure 1.1 Group table.

For groups of infinite order a group table is obviously impossible. Instead, the results of all possible compositions must be expressed by a general rule.

A concrete example of an abstract group, i.e., a group of concrete elements with a concretely defined law of composition, having the same structure as an abstract group, is called a *realization* of that abstract group. Such realizations might be, for example, groups of numbers, matrices, rotations, or other geometric transformations.

We now consider all abstract groups of orders 1 to 5 and one abstract group of order 6 and present examples, or realizations, for each but the last.

ORDER 1.   There is only one abstract group of order 1, denoted C_1, and it is the trivial group consisting only of the identity element e. It is Abelian. One realization of this group is the number 1 and ordinary multiplication as the composition. Another realization is the number 0 and the composition of ordinary addition.

ORDER 2.   There is only one abstract group of order 2, denoted C_2. It is Abelian. It consists of the identity element e and one other element a, which must then be its own inverse, $aa = e$. The group table is shown in Fig. 1.2. A realization of this group is the set of numbers $\{1, -1\}$ (the two square roots of 1) under ordinary multiplication. The number 1 serves as the identity, while the number -1 is its own inverse, $(-1) \times (-1) = 1$. Another, geometric, realization is the set consisting of the transformation of not doing anything, called the *identity transformation,* and the transformation of mirror reflection (with a two-sided mirror), the composition being consecutive reflection. (Transformations are discussed more thoroughly in Chapter 3.) The identity transformation serves as the group identity element, and the reflection transformation is its own inverse, since two consecutive reflections in the same mirror bring the situation back to the way it was originally. Another realization is the set of rotations about a common axis by $0°$, which is another way of expressing the identity transformation, and by $180°$, with the composition of consecutive rotation. Note that the group elements here are not the orientations of $0°$ and $180°$, but the *rotations* through those angles. Ro-

Figure 1.2   Group table of C_2.

e	a	b
a	b	e
b	e	a

Figure 1.3 Group table of C_3.

tation by 0° is the identity element. Rotation by 180° is its own inverse, since two consecutive rotations by 180° about the same axis give a total rotation by 360°—no rotation at all—the identity element.

ORDER 3. There is only one abstract group of this order, denoted C_3, and it is Abelian. It consists of the identity e and two more elements a and b, each of which is the inverse of the other,

$$ab = ba = e,$$

and each of which composed with itself gives the other,

$$aa = b, bb = a.$$

Figure 1.3 is the group table. One realization is the set of complex numbers $\{1, e^{2\pi i/3}, e^{4\pi i/3}\}$ (the three third roots of 1) under multiplication. The number 1 serves as the identity element, and the other two numbers can respectively correspond to either a, b or b, a. Another realization is the set of rotations about a common axis by $\{0°$ (the identity transformation), 120° $(= 360°/3)$, 240° $(= 2 \times 120°)\}$, with composition of consecutive rotation. The identity transformation of rotation by 0° corresponds to the identity element, while the other two rotations can respectively correspond to either a, b or b, a.

ORDER 4. There are two different abstract groups of order 4. Both are Abelian. One of them has a structure similar to that of the abstract group of order 3 and is denoted C_4. Its group table is shown in Fig. 1.4. It can

e	a	b	c
a	b	c	e
b	c	e	a
c	e	a	b

Figure 1.4 Group table of C_4.

e	a	b	c
a	e	c	b
b	c	e	a
c	b	a	e

Figure 1.5   Group table of D_2.

be realized by the set of numbers $\{1, i, -1, -i\}$ (the four fourth roots of 1) under multiplication. Another realization is the set of rotations about a common axis by $\{0°$ (the identity transformation), $90°$ $(= 360°/4)$, $180°$ $(= 2 \times 90°)$, $270°$ $(= 3 \times 90°)\}$, with consecutive rotation as composition.

The other group of order 4, denoted D_2, has the group table of Fig. 1.5. In this group each element is its own inverse,

$$aa = bb = cc = e.$$

To obtain a realization we imagine two perpendicular, intersecting two-sided mirrors A and B and their line of intersection C as in Fig. 1.6. The results of the transformations of reflection through each mirror and rotation by $180°$ about their line of intersection are shown in cross section in Fig. 1.7. The set of transformations consisting of these transformations and the identity transformation, with consecutive transformation as composition, forms a group and is a realization of D_2.

ORDER 5.   There is only one abstract group of order 5, denoted C_5, and it is Abelian. Its structure is similar to that of C_4. Figure 1.8 is its group

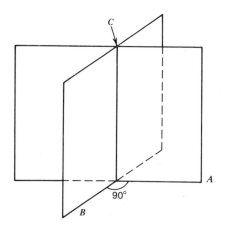

Figure 1.6.   Two perpendicular intersecting mirrors A and B and their line of intersection C.

Figure 1.7. Reflections in perpendicular mirrors A and B and rotation by $180°$ about their line of intersection C, shown in cross section.

table. One realization is the set of five fifth roots of 1 under multiplication. Another realization is the set of rotations $\{0°, 72° (= 360°/5), 144° (= 2 \times 72°), 216° (= 3 \times 72°), 288° (= 4 \times 72°)\}$, with consecutive rotation as composition.

All groups of orders 1 to 5 are Abelian. One might be tempted to guess by induction that all finite-order groups are Abelian. (This is the kind of "induction" by which one "proves" that 60 is divisible by all natural numbers: it is divisible by 2, by 3, by 4, by 5, by 6, and so on.) That is false. It turns out that the lowest-order non-Abelian group is one of the two order-6 groups and is denoted D_3. To dispel any lingering impression that all finite-order groups are Abelian, we display its group table in Fig. 1.9.

Now for some infinite-order groups. We present examples in terms of realizations, because the general rule for composition is most easily expressed in the context of a realization.

e	a	b	c	d
a	b	c	d	e
b	c	d	e	a
c	d	e	a	b
d	e	a	b	c

Figure 1.8 Group table of C_5.

e	a	b	c	d	f
a	b	e	f	c	d
b	e	a	d	f	c
c	d	f	e	a	b
d	f	c	b	e	a
f	c	d	a	b	e

Figure 1.9   Group table of D_3.

1.   The set of all integers (positive, negative, and zero) under addition forms a group. (1) Closure holds, since the sum of any two integers is an integer. (2) Addition of numbers is associative. (3) The identity of addition is the number 0, which is an integer. (4) The inverse of any integer a is its negative $-a$, which is also an integer. The group is Abelian, since addition is commutative. This set of numbers, however, does not form a group under ordinary multiplication. (1) Closure holds, since the product of any two integers is an integer. (2) Multiplication of numbers is associative. (3) The identity of multiplication is the number 1, which is an integer. But it is condition 4 that is not satisfied, since the multiplicative inverse of any integer a, which is its reciprocal $1/a$, is not in general an integer; for $a = 0$ it is not even defined.

2.   The set of all nonzero rational numbers does form a group under multiplication. (1) The product of any two rational numbers is rational. (2) Multiplication is associative and even commutative, so the group is Abelian. (3) The identity of multiplication, the number 1, is rational. (4) The multiplicative inverse of any rational number a ($\neq 0$) is its reciprocal $1/a$, which is also rational, since if a is the ratio of two integers, so is $1/a$. (This set together with the number 0 forms a group under addition.)

3.   The set of all $n \times n$ matrices forms a group under matrix addition. (1) The sum of any two $n \times n$ matrices is an $n \times n$ matrix. (2) Matrix addition is associative. It is also commutative, so the group is Abelian. (3) The identity of addition is the zero $n \times n$ matrix, which is a member of the set. (4) The inverse of any $n \times n$ matrix is its negative, which is also an $n \times n$ matrix. The same set of matrices fails to form a group under matrix multiplication. It has trouble with condition 4.

4.   However, the set of all *nonsingular* $n \times n$ matrices does form a group under matrix multiplication. (1) The product of any two nonsingular $n \times n$ matrices is a nonsingular $n \times n$ matrix. (2) Matrix multiplication is associative. It is not commutative, so the group is non-Abelian (except for $n = 1$). (3) The identity of matrix multiplication is the unit $n \times n$

matrix, which is nonsingular. (4) The inverse of any nonsingular $n \times n$ matrix is its matrix inverse, which exists and is a nonsingular $n \times n$ matrix. This set of matrices does not form a group under addition, since it does not fulfill conditions 1 and 3.

5.   The set of all real orthogonal $n \times n$ matrices under matrix multiplication forms a group. (1) The product of any two real orthogonal $n \times n$ matrices is a real orthogonal $n \times n$ matrix. (2) Matrix multiplication is associative. (3) The unit $n \times n$ matrix, the identity of matrix multiplication, is real and orthogonal. (4) The inverse of any real orthogonal $n \times n$ matrix is its matrix inverse, which exists and is also a real orthogonal $n \times n$ matrix. This group is non-Abelian for $n > 2$.

6.   The set of all displacements (also called translations) in a common direction forms a group with the composition of consecutive displacement. (1) The composition of any two displacements in a common direction is a displacement in the same direction. In fact, the composition of displacement by a cm and displacement by b cm in a common direction is displacement by $(a + b)$ cm in the same direction. We see that the group is Abelian. (2) Composition by consecutive displacement is associative. This might or might not be clear. In Chapter 3 we see that composition of transformations by consecutive transformation is always associative. (3) The identity transformation, the nonaction of doing nothing (just wonderful for lazy people), is a member of the set, since the null displacement, displacement by 0 cm, is the identity transformation. The identity transformation is the identity of any group of transformations with consecutive transformation as composition. This should be clear but will be discussed in detail in Chapter 3. (4) For any displacement the inverse is also a displacement in the same direction. It is displacement by the same distance, in the same direction, but in the opposite sense. In other words, the inverse of displacement by a cm is displacement by $-a$ cm in the same direction.

7.   The set of all rotations about a common axis, with composition of consecutive rotation, forms a group. (1) The composition of any two rotations about the same axis is a rotation about the same axis. Rotation by $a°$ followed by rotation by $b°$ about the same axis results in total rotation by $(a + b)°$ about the same axis. The group is clearly Abelian. (2) Composition by consecutive rotation is associative. This is discussed in Chapter 3. (3) The identity transformation is a member of the set, since the null rotation, rotation by $0°$, is the identity transformation. (4) The inverse of any rotation is also a rotation about the same axis. For a rotation by $a°$ the inverse might be expressed as rotation by $-a°$ about the same axis or as rotation by $(360 - a)°$ about the same axis.

8.   The set of all rotations about a point, consisting of all rotations about all axes through a point, forms a group under composition of consecutive rotation. (1) Closure is far from obvious, and we do not show it here. (2) Associativity is discussed in Chapter 3 for transformations in general. (3) The identity transformation is a member of the set; it is the null rotation about any axis through the point. (4) The inverse of any rotation by $a°$ about any axis through the point is rotation by $-a°$ about the same axis and is thus also a member of the set. The group is non-Abelian. To show this consider Fig. 1.10, in which rotation by 90° about the axis through the center of the square and perpendicular to the plane of the paper and rotation by 180° about axis AB through the center of the square are applied to the marked square consecutively in each possible order. The two results are different. Since at least these two rotations, both members of the group, are noncommuting, the group is non-Abelian.

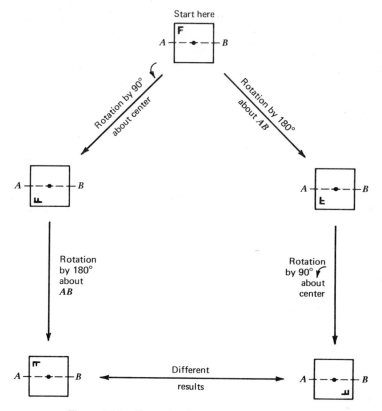

Figure 1.10.   Example of noncommuting rotations.

Problems

1. (a) Prove that in a finite-order group for every element g there exists a positive integer k such that $g^k = e$. (*Hint:* Consider the sequence $g, g^2, g^3, \ldots$ and remember that the order of the group is finite.) The smallest such k is called the *order* (or *period*) of element g. (b) Find the order of each element of the order-6 group D_3.

2. If a, b, and ab are group elements of order 2, prove that a and b commute. (See preceding problem.)

3. Prove that each set of matrices forms a group under matrix multiplication by constructing its group table and identifying the order-4 group of which it is a realization:

    (a) $\left\{ \begin{pmatrix} 1 & 0 \\ 0 & 1 \end{pmatrix}, \begin{pmatrix} 0 & 1 \\ -1 & 0 \end{pmatrix}, \begin{pmatrix} -1 & 0 \\ 0 & -1 \end{pmatrix}, \begin{pmatrix} 0 & -1 \\ 1 & 0 \end{pmatrix} \right\}$;

    (b) $\left\{ \begin{pmatrix} 1 & 0 \\ 0 & 1 \end{pmatrix}, \begin{pmatrix} 1 & 0 \\ 0 & -1 \end{pmatrix}, \begin{pmatrix} -1 & 0 \\ 0 & 1 \end{pmatrix}, \begin{pmatrix} -1 & 0 \\ 0 & -1 \end{pmatrix} \right\}$.

4. Prove that the identity transformation and rotations by 180° about each of three mutually perpendicular, intersecting straight lines with a common point of intersection form an order-4 group under composition of consecutive transformation. Construct the group table. Of which abstract order-4 group is this a realization?

5. Prove that the set of matrices

    $$\left\{ \begin{pmatrix} 1 & 0 \\ 0 & 1 \end{pmatrix}, \begin{pmatrix} \omega & 0 \\ 0 & \omega^2 \end{pmatrix}, \begin{pmatrix} \omega^2 & 0 \\ 0 & \omega \end{pmatrix}, \begin{pmatrix} 0 & 1 \\ 1 & 0 \end{pmatrix}, \begin{pmatrix} 0 & \omega^2 \\ \omega & 0 \end{pmatrix}, \begin{pmatrix} 0 & \omega \\ \omega^2 & 0 \end{pmatrix} \right\},$$

    where $\omega^3 = 1$ but $\omega \neq 1$, forms a group under matrix multiplication, and prove that it is a realization of the group D_3.

6. The law of composition of the set of functions

    $$\left\{ f_1(x) = x, \quad f_2(x) = \frac{1}{1-x}, \quad f_3(x) = \frac{x-1}{x}, \right.$$
    $$\left. f_4(x) = \frac{1}{x}, \quad f_5(x) = 1-x, \quad f_6(x) = \frac{x}{x-1} \right\}$$

    is substitution. For example,

    $$f_4 f_3 = f_4(f_3(x)) = \frac{1}{f_3(x)} = \frac{x}{x-1} = f_6.$$

    Prove that this set forms a group under this composition by constructing its group table, and show that it is a realization of the group D_3.

7. Why does not each set form a group under the specified law of composition (check all four conditions)? (a) All odd integers under addition; (b) $\{\frac{1}{2}, \frac{1}{3}, 1, 2, 3\}$ under multiplication; (c) all positive rational numbers, where the composition of a and b is a/b; (d) all ordinary vectors (3-vectors) under vector product.

8. Prove that each set of numbers forms an infinite-order (Abelian) group under ordinary multiplication: (a) $\{2^k\}$ for integer k; (b) $\{(1 + 2m)/(1 + 2n)\}$ for integer m and n; (c) $\{a + b\sqrt{5}\}$ for rational a and b such that $a^2 + b^2 \neq 0$; (d) $\{\cos\alpha + i \sin\alpha\}$ for real α.

9. Each of the following sets of numbers forms an infinite-order (Abelian) group either under addition or under multiplication. For each set check which is the group composition and prove that the set is indeed a group: (a) all real numbers; (b) all nonzero real numbers; (c) all positive reals; (d) all complex numbers; (e) all nonzero complex numbers; (f) all unimodular complex numbers (i.e., all complex numbers whose absolute value is 1).

10. (a) Prove that the set of all real polynomials in one variable forms a group under addition. (b) Why does the set of all nonzero real polynomials in one variable not form a group under multiplication?

11. (a) Prove that the set of all unitary $n \times n$ matrices forms a group under matrix multiplication. (b) Prove that the set of all unimodular (determinant $= 1$) real orthogonal $n \times n$ matrices forms a group under matrix multiplication.

12. Prove that the set of matrices

$$L(v) = (1 - v^2/c^2)^{-1/2} \begin{pmatrix} 1 & -v \\ -v/c^2 & 1 \end{pmatrix}$$

for all v in the range $-c < v < c$, with c a positive constant, forms an Abelian "continuous" group in the sense that

$$L(v_1)L(v_2) = L(v_3),$$

where

$$v_3 = \frac{v_1 + v_2}{1 + v_1 v_2/c^2}.$$

This is the group of Lorentz transformations in the same direction, where c is the speed of light.

13. Prove that the set of all matrices of the form $\begin{pmatrix} a & -b \\ b & a \end{pmatrix}$ with a and b real forms an Abelian group under matrix multiplication as long as a and b satisfy a certain condition. What is this condition?

1.2. MAPPING

At this point I think it is worthwhile to devote a short discussion to the concept of *mapping,* because we make much use of mappings in the rest of the book. A mapping is a correspondence made from one set to another (or to the same set). (The sets may or may not be groups.) A mapping from set A to set B puts every element of A in correspondence with some element of B, as in Fig. 1.11. A mapping is denoted $A \rightarrow B$. If the mapping is given a name, say the mapping M from A to B, it is denoted $A \xrightarrow{M} B$ or $M: A \rightarrow B$.

An element of B that is in correspondence with an element of A is called the *image* of the element of A. The element of A is called an *object* of the element of B. A mapping $A \rightarrow B$ such that an element of B may be the image of more than one element of A is called a *many-to-one* mapping. If, however, different elements of A are always in correspondence with different elements of B, i.e., if every image in B is the image of a unique object in A, the mapping is called *one-to-one*. (A one-to-one mapping is also called an *injective* mapping.) Figure 1.12 illustrates a one-to-one mapping.

If not all elements of B are necessarily images under the mapping, it is called a mapping of A *into* B. Figures 1.11 and 1.12 illustrate such

Figure 1.11. Many-to-one mapping of A into B.

Figure 1.12. One-to-one mapping of A into B.

mappings. If, however, every element of **B** is the image of at least one element of **A**, the mapping is from **A** *onto* **B**, as in Fig. 1.13. (An onto mapping is also called a *surjective* mapping.)

A mapping may also be denoted in terms of the set elements, as $a \rightarrow b$ or $a \xrightarrow{M} b$, where a represents elements of **A** and b represents elements of **B**. We use this notation extensively. Another notation, using the notation commonly used for functions, is $b = M(a)$. We also use this notation, but not in Chapters 1 and 2.

A mapping may be from a set to itself, i.e., **A** and **B** may be the same set. Familiar examples of such mappings are real functions,

$$x \rightarrow y = f(x),$$

where the set is the set of all real numbers. The function $y = x^2$ is a

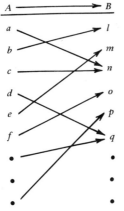

Figure 1.13. Many-to-one mapping of A onto B.

many-to-one, actually two-to-one except for $y = 0$, mapping of the reals into the reals (only the nonnegative numbers are images). The function $y = e^x$ is a one-to-one mapping of the reals into the reals (only positive numbers are images). The function $y = \sin x$ is an infinity-to-one mapping of the reals into the reals (only numbers whose absolute values are not larger than 1 are images). The function $y = 3x - 7$ is a one-to-one mapping from the reals onto the reals.

A one-to-one mapping from A onto B may be inverted, simply by reversing the sense of the correspondence arrows, to obtain a one-to-one mapping from B onto A. The latter mapping is called the *inverse* of the former. All elements of B are objects under the inverse mapping, because all elements of B are images for the direct mapping. The inverse mapping is one-to-one, since the direct mapping is one-to-one. And the inverse mapping is onto, because all elements of A are objects under the direct mapping. Thus the inverse mapping is indeed a one-to-one mapping from B onto A, as in Fig. 1.14. Its inverse is the direct mapping itself. If the direct mapping is denoted M, its inverse is denoted M^{-1}, so that $A \xrightarrow{M} B$, $B \xrightarrow{M^{-1}} A$, $a \xrightarrow{M} b$, $b \xrightarrow{M^{-1}} a$, $b = M(a)$, $a = M^{-1}(b)$. Note that only a mapping that is both one-to-one and onto has an inverse; none of the other kinds of mapping just discussed is invertible. (A mapping that is both one-to-one and onto, i.e., is both injective and surjective, is called *bijective*. Thus bijectiveness and invertibility imply one another.)

As an example of an inverse mapping, taking the one-to-one mapping

$$x \to y = 3x - 7$$

from the reals onto the reals, the inverse mapping is

$$y \to x = (y + 7)/3.$$

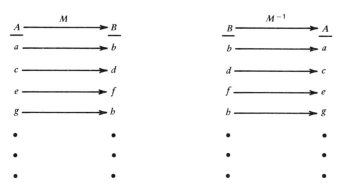

Figure 1.14.   One-to-one mapping M of A onto B and its inverse M^{-1}.

Problems

1. Each of the following is a mapping of the set of all points of a plane to itself. Prove that each mapping is one-to-one and onto so that it has an inverse. Find the inverse. (a) The image of any point is found by moving from it 3 units in the x direction and 2 units in the y direction; (b) the image of any point is found by moving from it counterclockwise around the circle through the point with center at the origin; (c) the image of any point is found by moving from it along the straight line through the point and the origin, away from the origin to twice the distance from the origin.

2. Consider the mapping from the set of all complex numbers to the set of all real numbers: $z \rightarrow |z|$. Is this mapping one-to-one or many-to-one? Is it into or onto? Is it invertible? If we considered it as a mapping from the set of all complex numbers to the set of all non-negative numbers, how would you answer these questions?

3. For each of the following mappings answer the questions: Is the mapping one-to-one or many-to-one, and, if the latter, how many? Is the mapping into or onto? Is it an invertible mapping? If so, what is the inverse mapping? If not, why not? (a) Mapping by perpendicular projection of the set of all points of three-dimensional space to the set of all points of a plane in that space, i.e., the image in the plane of any point in space is found by dropping a perpendicular from the point to the plane; (b) mapping by perpendicular projection of the set of all points of a straight line to the set of all points of a plane that is not perpendicular to the line, i.e., the image of any point on the line is found by dropping a perpendicular from the point to the plane; (c) mapping by perpendicular projection of the set of all points of a straight line to the set of all points of another straight line that is not orthogonal to the first line, i.e., the image of any point on the first line is found by dropping a perpendicular from the point to the second line.

4. For each of the following functions $y = f(x)$ answer the questions: Does

$$x \rightarrow y = f(x)$$

define a mapping $R \rightarrow R'$, where both R and R' denote the set of all real numbers? If not, can R be reduced (e.g., to the set of all positive numbers) to make a mapping? Is the mapping one-to-one or many-to-one (how many?)? If it is many-to-one, can R be reduced (further) to make the mapping one-to-one? Is the mapping into or onto? If it is into, reduce R' to make the mapping onto. If you have obtained a one-to-one mapping that is onto, it is invertible. What is the corresponding inverse function? (a) $y = x^2$; (b) $y = e^x$; (c) $y = \sin x$; (d) $y = \tan x$; (e) $y = +\sqrt{x}$; (f) $y = \ln x$; (g) $y = 2x + \sqrt{5}$.

5. Consider a circle, a straight line tangent to it, and a point P located anywhere in the plane of the circle except on the straight line. Denote the set of all points on the circle by C and the set of all points on the line by L. We define the correspondence between points of C and points of L: A point of C and a point of L are in correspondence if and only if they are collinear with P, i.e., if the straight line through them passes also through P. Consider this correspondence for each of the following locations of P and answer the questions: Does the correspondence define a mapping $C \to L$ or $L \to C$ or no such mapping at all? If it defines such a mapping, is it one-to-one, many-to-one (how many?), into, onto? Is it an invertible mapping? (a) The line separates P and the circle; (b) P is on the same side of the line as the circle, but outside the circle; (c) P is inside the circle; (d) P is on the circle, but not diametrically opposite the point of tangency; (e) P is on the circle, diametrically opposite the point of tangency. (*Note*: The question "How many?" might not have the same answer for all images of a mapping. To be able to set up a mapping, you might find it necessary to remove one or more points from C or, alternatively, to add to L the "point at infinity.")

1.3. ISOMORPHISM

Consider a many-to-one or one-to-one mapping of a group G onto another group G', so that every element of G has an image in G' and every element of G' is an image of at least one element of G, as in Fig. 1.15.

Now consider such a mapping that preserves structure. What we mean by this is that, if a and b are any two elements of G with images a' and b', respectively, in G', and if we denote by $(ab)'$ the image in G' of the composition ab in G, then

$$a'b' = (ab)'$$

(where the composition $a'b'$ is, of course, in G'). In other words, the image of a composition is the composition of images. This can also be

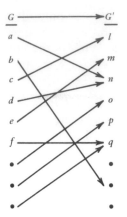

Figure 1.15. Many-to-one mapping of G onto G'.

expressed by the diagram for all a, b in G:

$$
\begin{array}{ccc}
a & b & = c \\
\downarrow & \downarrow & \downarrow \\
a' & b' & = c'.
\end{array}
$$

Such a structure-preserving mapping is called a *homomorphism*. We postpone the discussion of homomorphism in general to Section 1.5.

If a homomorphism is one-to-one, i.e., if each element of G' is the image of exactly one element of G so that the mapping can be inverted, it is called an *isomorphism* and is denoted $G \sim G'$. Figure 1.16 shows such a mapping. And the structure-preserving property of an isomorphism

$$
\begin{array}{cc}
G & \sim & G' \\
a & \longleftrightarrow & a' \\
b & \longleftrightarrow & b' \\
c & \longleftrightarrow & c' \\
d & \longleftrightarrow & d' \\
e & \longleftrightarrow & e' \\
f & \longleftrightarrow & f' \\
\bullet & & \bullet \\
\bullet & & \bullet \\
\bullet & & \bullet
\end{array}
$$

Figure 1.16. Isomorphism mapping of G and G'.

can be expressed by the diagram

$$\begin{array}{ccc} a & b & = & c \\ \updownarrow & \updownarrow & & \updownarrow \\ a' & b' & = & c'. \end{array}$$

Isomorphic groups have the same structure. And groups having the same structure, in the sense we used previously, that their group tables can be made the same by reordering rows and columns and redenoting elements, are isomorphic. So "isomorphic" makes precise the concept of "having the same structure" and is more general, since it is applicable to infinite-order as well as finite-order groups. Thus all realizations of an abstract group are isomorphic with the abstract group of which they are realizations as well as with each other.

Consider some examples of isomorphism.

1.   The abstract group of order 3 C_3 and two of its realizations, by rotations and by complex numbers. See Fig. 1.17.

2.   The Abelian infinite-order group of integers under addition is isomorphic with the Abelian infinite-order group of integral powers of 2 (or of any other positive number) under multiplication. The mapping is $n \leftrightarrow 2^n$. The identities are $0 \leftrightarrow 2^0 = 1$. Structure is preserved by the mapping:

$$\begin{array}{ccc} m & + & n & = & (m + n) \\ \updownarrow & & \updownarrow & & \updownarrow \\ 2^m & \times & 2^n & = & 2^{m+n}. \end{array}$$

3.   The Abelian infinite-order group of all real numbers under addition is isomorphic with the Abelian infinite-order group of all displacements

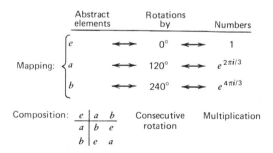

Figure 1.17.   Isomorphism of C_3 and two of its realizations.

in a common direction with the composition of consecutive displacement. The mapping is (number a) $\leftrightarrow$ (displacement by a cm). See Fig. 1.18 for structure preservation.

Here and in the following we adopt the convention that consecutive transformations are read from right to left. The reason will be made clear in Chapter 3. For the time being we use " $+$ " to denote the composition of consecutive transformation.

4.   The Abelian infinite-order group of all rotations about a common axis, with composition of consecutive rotation, is isomorphic with the Abelian infinite-order group of all unimodular (determinant $=$ 1) real orthogonal 2 $\times$ 2 matrices under matrix multiplication. Both are isomorphic with the group of unimodular (absolute value $=$ 1) complex numbers under multiplication. Note that the most general 2 $\times$ 2 matrix that is unimodular, real, and orthogonal is $\begin{pmatrix} \cos\alpha & -\sin\alpha \\ \sin\alpha & \cos\alpha \end{pmatrix}$ with α real. The most general unimodular complex number is $e^{i\alpha}$ with α real. The mapping for these isomorphisms is

$$\text{(rotation by angle } \alpha) \leftrightarrow \begin{pmatrix} \cos\alpha & -\sin\alpha \\ \sin\alpha & \cos\alpha \end{pmatrix} \leftrightarrow e^{i\alpha}.$$

The identities are

$$\text{(rotation by angle 0)} \leftrightarrow \begin{pmatrix} 1 & 0 \\ 0 & 1 \end{pmatrix} \leftrightarrow e^0 = 1.$$

For structure preservation see Fig. 1.19.

5.   The non-Abelian infinite-order group of all rotations about a common point under consecutive rotation is isomorphic with the non-Abelian infinite-order group of unimodular real orthogonal 3 $\times$ 3 matrices under

Figure 1.18.   Structure preservation for isomorphism of group of real numbers under addition with group of displacements in common direction (with composition of consecutive displacement, denoted " $+$ ").

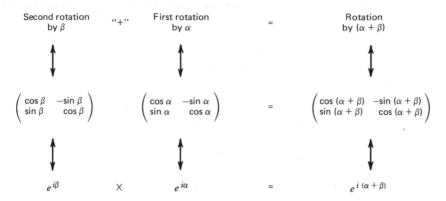

| Second rotation
by β | "+" | First rotation
by α | = | Rotation
by $(\alpha + \beta)$ |

$$\begin{pmatrix} \cos \beta & -\sin \beta \\ \sin \beta & \cos \beta \end{pmatrix} \qquad \begin{pmatrix} \cos \alpha & -\sin \alpha \\ \sin \alpha & \cos \alpha \end{pmatrix} \quad = \quad \begin{pmatrix} \cos (\alpha + \beta) & -\sin (\alpha + \beta) \\ \sin (\alpha + \beta) & \cos (\alpha + \beta) \end{pmatrix}$$

$$e^{i\beta} \qquad \times \qquad e^{i\alpha} \qquad = \qquad e^{i(\alpha + \beta)}$$

Figure 1.19. Structure preservation for isomorphism of group of rotations about common axis (with composition of consecutive rotation, denoted "+"), group of unimodular real orthogonal 2×2 matrices under matrix multiplication, and group of unimodular complex numbers under multiplication.

matrix multiplication. The mapping is (rotation by angle α about axis with direction cosines (λ,μ,ν)) $\leftrightarrow$ (a certain unimodular real orthogonal 3×3 matrix $M(\lambda,\mu,\nu;\alpha)$ uniquely determined by (λ,μ,ν) and α in a way that we do not go into here). See [M34] and [M33] in the Bibliography. For structure preservation see Fig. 1.20.

Problems

1. Prove that groups of different finite orders cannot be isomorphic with each other.

2. Prove that in an isomorphism corresponding elements have the same order. (See Problem 1 of Section 1.1.)

3. Prove that a group is Abelian if and only if the correspondence

| Second rotation
by α_2 about
$(\lambda_2, \mu_2, \nu_2)$ | "+" | First rotation
by α_1 about
$(\lambda_1, \mu_1, \nu_1)$ | = | Rotation
by α_3 about
$(\lambda_3, \mu_3, \nu_3)$ |

$$M(\lambda_2, \mu_2, \nu_2; \alpha_2) \qquad M(\lambda_1, \mu_1, \nu_1; \alpha_1) \quad = \quad M(\lambda_3, \mu_3, \nu_3; \alpha_3)$$

Figure 1.20. Structure preservation for isomorphism of group of rotations about common point (with composition of consecutive rotation, denoted "+") with group of unimodular real orthogonal 3×3 matrices $M(\lambda,\mu,\nu;\alpha)$ under matrix multiplication.

$a \leftrightarrow a^{-1}, b \leftrightarrow b^{-1}, \ldots$ for all its elements is an isomorphism (of the group with itself).

4. All the previous examples and problems formulated in terms of realizations are, of course, isomorphism examples and problems. Formulate them rigorously in terms of isomorphism.

5. Prove that the group of all positive numbers under multiplication is isomorphic with the group of all real numbers under addition. (*Hint:* Consider the function $y = e^x$ or its inverse, $x = \ln y$. This isomorphism is the theoretical basis for the use of logarithms or a slide rule for calculating products, ratios, powers, and roots. However, by now the pocket calculator has superseded logarithms and slide rules.)

6. Prove that C_n, the group of rotations about a common axis by $\{0°, 360°/n, 2 \times 360°/n, 3 \times 360°/n, \ldots, (n - 1) \times 360°/n\}$, with composition of consecutive rotation, is isomorphic with the group consisting of the n nth roots of 1, $\{1, e^{2\pi i/n}, e^{2 \times 2\pi i/n}, e^{3 \times 2\pi i/n}, \ldots, e^{(n-1) \times 2\pi i/n}\}$, under multiplication. (See the first example in this section.)

7. Prove that the group of matrices of the form $\begin{pmatrix} a & -b \\ b & a \end{pmatrix}$, with a and b real and obeying a certain condition, under matrix multiplication is isomorphic with the group of nonzero complex numbers under multiplication. (See Problem 13 of Section 1.1.)

1.4. EQUIVALENCE RELATION

At this most appropriate point we take the opportunity to introduce the concept of *equivalence relation,* of which isomorphism is an example, as we will see. An equivalence relation is defined as any relation, denoted here by $\equiv$, that holds between the members of pairs of elements of a set and satisfies the following three conditions:

1. *Reflexivity.* Every element of the set has the relation with itself, i.e.,

$$a \equiv a$$

for all a.

2. *Symmetry.* If one element has the relation with another, then the second has it with the first and vice versa, for all elements of the set. In

symbols this is

$$a \equiv b \Leftrightarrow b \equiv a$$

for all a, b. (The arrow $\Rightarrow$ denotes implication in the direction of the arrow. The double arrow $\Leftrightarrow$ denotes implication both ways. Recall that whatever stands at the head of an implication arrow is a necessary condition for whatever stands at the tail. And whatever stands at the tail is a sufficient condition for whatever stands at the head. Whatever stands at one head of a double implication arrow is a necessary and sufficient condition for whatever stands at the other head.)

3. *Transitivity.* If one element has the relation with a second and the second has it with a third, then the first element has the relation with the third, for all elements of the set. Represented symbolically this is

$$a \equiv b, \, b \equiv c \Rightarrow a \equiv c$$

for all a, b, c.

One somewhat fanciful example of an equivalence relation is friendship. (1) It is a reflexive relation, since everyone is a friend of himself or herself (well, we hope so, anyhow). (2) It is symmetric, since if you are my friend I am your friend. (3) And it is transitive, since a friend of a friend is a friend (again, this is the ideal).

More seriously, the most familiar example of equivalence relation is equality, as denoted by $=$. This hardly needs explaining.

Isomorphism is an equivalence relation. The elements here are groups, members of the set of all groups (and not the elements of the groups!). (1) Isomorphism is reflexive, because every group is isomorphic with itself. The mapping is just every element is its own image. See Fig. 1.21. (2) Isomorphism is symmetric, since the mapping is one-to-one and onto,

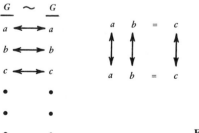

Figure 1.21. Reflexivity of isomorphism.

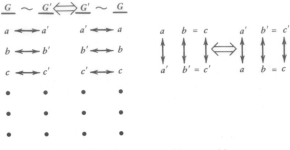

Figure 1.22. Symmetry of isomorphism.

thus invertible. See Fig. 1.22. (3) Isomorphism is transitive. Figure 1.23 is worth a hundred words.

The congruence relation of modular arithmetic, "$\equiv$ (mod n)," is an equivalence relation.

An example of a relation that is not an equivalence relation is the relation of "lesser than," denoted $<$. It is not reflexive, since no number is lesser than itself. Neither is it symmetric, since, if a is lesser than b, certainly b is not lesser than a. It is, however, transitive; if a is lesser than b and b is lesser than c, then a is lesser than c.

Another example is the relation of inequality, denoted $\neq$. This relation is symmetric but not reflexive or transitive. (If a does not equal b and b does not equal c, nothing is implied about the relation between a and c.)

Intimately related to the concept of equivalence relation is the concept of *equivalence class*, but we postpone presenting it until the next chapter.

Problems

1. You might think that the reflexivity property of an equivalence relation is redundant in that it follows from the other two properties. Show that, if you can. However, a certain assumption is being made.

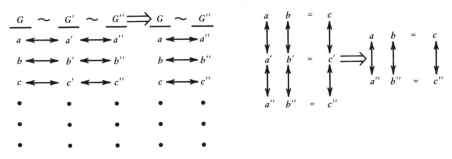

Figure 1.23. Transitivity of isomorphism.

Which? (What about reflexivity for an element that is not equivalent to any other element?)

2. Prove that the congruence relation of modular arithmetic, "$\equiv$ (mod n)," is an equivalence relation.

3. Check whether each of the following relations is an equivalence relation or not. If not, for which of the three defining properties does it fail? (a) The relation $\geq$ among real numbers; (b) having the same absolute value for complex numbers; (c) having the same value of determinant for matrices; (d) parallelism among straight lines in three-dimensional space; (e) orthogonality among straight lines in three-dimensional space; (f) having at least one root in common for quadratic equations.

1.5. HOMOMORPHISM

We now return to the discussion of homomorphism (of which isomorphism is a special case) in general, which we abandoned in Section 1.3. Such a mapping is not invertible (unless it is an isomorphism and only then). Homomorphism is denoted $G \to G'$, where the mapping is many-to-one from G to G', as in Fig. 1.24. The structure-preserving property of homomorphism can be expressed by the diagram

$$\begin{array}{ccc} a & b & = & c \\ \downarrow & \downarrow & & \downarrow \\ a' & b' & = & c'. \end{array}$$

Figure 1.24. Homomorphism mapping of G to G'.

Although homomorphism preserves structure in the sense just shown, homomorphic groups do not have the same structure, unless they are isomorphic. For example, non-Abelian groups may be homomorphic to Abelian groups (the opposite is impossible, however). Infinite-order groups may be homomorphic to finite-order groups, and groups of different finite orders may be homomorphic. Examples are presented shortly. Homomorphism is not an equivalence relation, since it is not symmetric. It is reflexive, however, with isomorphism as a special case of homomorphism. See Fig. 1.21. And it is transitive. See Fig. 1.25.

In a homomorphism $G \rightarrow G'$ that set of all elements of G whose images are the identity element of G' is called the *kernel* of the homomorphism. We will return to this in Section 2.1.

Some examples of homomorphism follow.

1. Any group whatsoever is trivially homomorphic to the abstract group of order 1 C_1, as in Fig. 1.26. Structure is preserved:

$$\begin{array}{ccc} a & b & = c \\ \downarrow & \downarrow & \downarrow \\ e' & e' & = e'. \end{array}$$

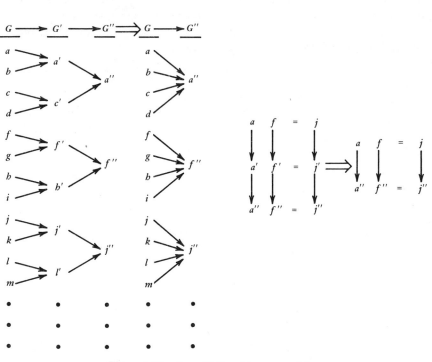

Figure 1.25. Transitivity of homomorphism.

$$G \longrightarrow G'$$

Figure 1.26.   Trivial homomorphism of any group to order-1 group.

All of G is the kernel of such a homomorphism. In terms of realizations, we can homomorphically map all the elements of any group to the number 1 (under multiplication) or to the number 0 (under addition).

   2.   Both abstract groups of order 4 (or any realizations thereof) are homomorphic to the abstract group of order 2 C_2 (or any of its realizations). For C_4 the mapping is shown in Fig. 1.27. For D_2 there are three possible mappings, shown in Fig. 1.28. Structure preservation can be checked with the help of the group tables (Section 1.1). The kernel of the first homomorphism (Fig. 1.27) is $\{e,b\}$. The three possible kernels of the second homomorphism (Fig. 1.28) are $\{e,a\}$, $\{e,b\}$, $\{e,c\}$.

   3.   The non-Abelian order-6 group D_3 is homomorphic to Abelian C_2. The mapping is shown in Fig. 1.29. Preservation of structure is checked with the group tables (Section 1.1). The kernel of this homomorphism is $\{e,a,b\}$.

   4.   The Abelian infinite-order group of nonzero real numbers under multiplication is homomorphic to Abelian C_2 or its realization, for example, by the numbers $\{1, -1\}$ under multiplication. The mapping is

$$\text{any positive number} \rightarrow e \leftrightarrow 1,$$
$$\text{any negative number} \rightarrow a \leftrightarrow -1.$$

$$C_4 \longrightarrow C_2$$

Figure 1.27.   Homomorphism of C_4 to C_2.

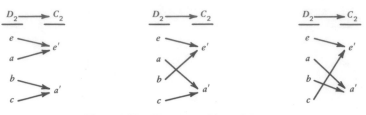

Figure 1.28.   Homomorphism of D_2 to C_2.

Structure preservation is seen as follows:

$$
\begin{array}{ccc}
\text{positive} & \times & \text{positive} & = & \text{positive} \\
\text{number} & & \text{number} & & \text{number} \\
\downarrow & & \downarrow & & \downarrow \\
e & & e & = & e \\
\updownarrow & & \updownarrow & & \updownarrow \\
1 & \times & 1 & = & 1
\end{array}
$$

$$
\begin{array}{ccc}
\text{positive} & \times & \text{negative} & = & \text{negative} \\
\text{number} & & \text{number} & & \text{number} \\
\downarrow & & \downarrow & & \downarrow \\
e & & a & = & a \\
\updownarrow & & \updownarrow & & \updownarrow \\
1 & \times & (-1) & = & (-1)
\end{array}
$$

$$
\begin{array}{ccc}
\text{negative} & \times & \text{positive} & = & \text{negative} \\
\text{number} & & \text{number} & & \text{number} \\
\downarrow & & \downarrow & & \downarrow \\
a & & e & = & a \\
\updownarrow & & \updownarrow & & \updownarrow \\
(-1) & \times & 1 & = & (-1)
\end{array}
$$

$$
\begin{array}{ccc}
\text{negative} & \times & \text{negative} & = & \text{positive} \\
\text{number} & & \text{number} & & \text{number} \\
\downarrow & & \downarrow & & \downarrow \\
a & & a & = & e \\
\updownarrow & & \updownarrow & & \updownarrow \\
(-1) & \times & (-1) & = & 1
\end{array}
$$

The set of all positive real numbers comprises the kernel of this homomorphism.

5.   The non-Abelian (for $n > 1$) group of all nonsingular complex $n \times n$ matrices under matrix multiplication is homomorphic to the Abe-

lian group of nonzero complex numbers under multiplication. Each matrix A is mapped to the complex number which is the value of its determinant, $A \rightarrow |A|$ (so that $A^{-1} \rightarrow 1/|A|$). Structure preservation follows from the fact that the determinant of a matrix product is the product of the individual determinants:

$$
\begin{array}{ccc}
A & B & = (AB) \\
\downarrow & \downarrow & \downarrow \\
|A| \times |B| & = & |AB|
\end{array}
$$

The unimodular $n \times n$ matrices comprise the kernel of this homomorphism, since they are all the matrices whose image is 1, the identity of number multiplication.

Problems

1. Prove that an Abelian group cannot be homomorphic to a non-Abelian group.

2. Prove that the group of nonzero real numbers under multiplication is homomorphic to the group of positive numbers under multiplication. What is the kernel?

3. (a) Prove that the group of nonzero complex numbers under multiplication is homomorphic to the group of positive numbers under multiplication. What is the kernel? (b) Find two different homomorphisms of the group of complex numbers under addition to the group of real numbers under addition. What are the kernels?

4. Prove that the group of nonsingular complex $n \times n$ matrices under matrix multiplication is homomorphic to the group of unimodular complex $n \times n$ matrices under matrix multiplication. (*Hint:* Consider the mapping: $A \rightarrow A/|A|^{1/n}$.) What is the kernel?

Figure 1.29. Homomorphism of D_3 to C_2.

5. Prove that the operation of differentiation maps the group of all po-
    lynomials in one variable (under addition) onto itself and prove that
    this mapping is a homomorphism. What is the kernel?

1.6. SUBGROUP

A *subset* of a set A is any set all of whose elements are elements of A.
For example, the set of natural numbers is a subset of the set of integers,
or the set of real orthogonal $n \times n$ matrices is a subset of the set of
unitary $n \times n$ matrices. Note that according to this definition any set is
a subset of itself. If B is a subset of A, this relation is denoted $B \subset A$ or
$A \supset B$, where $\subset$ and $\supset$ denote the relation of *inclusion*. (Sometimes $\subset$
and $\supset$ are used to denote only strict inclusion, where the sets may not
be the same, so that there is at least one element in the including set
that is not a member of the subset. Inclusion that allows the subset to
coincide with the including set is then denoted $\subseteq$ or $\supseteq$. However, we do
not use this convention.)

   If a subset of the elements of a group G itself forms a group with
respect to the composition law of G, it is called a *subgroup* of G. If H
is a subgroup of G, the relation is denoted $H \subset G$ or $G \supset H$, using the
same notation used for set inclusion.

   Every group G has two trivial subgroups: (1) the group of order 1
consisting of the identity element of G and (2) the group G itself. All the
subgroups of a group except the group itself are called *proper subgroups*.

   It can be shown, and indeed we will show in Section 2.2, that if a
finite-order group of order n includes a subgroup of order m, then m is
a divisor of n, i.e., $n = ms$ for some integer s. From this theorem it
follows that groups of prime order have no proper subgroups except the
trivial subgroup of order 1. Thus the groups of orders 2, 3, and 5, which
were presented in Section 1.1, can have no nontrivial proper subgroups,
as an examination of their group tables can confirm. In any case the group
of order 2 is of too low an order to have nontrivial proper subgroups.

   If G_2 is a subgroup of G_1 and G_3 a subgroup of G_2, then clearly G_3 is
also a subgroup of G_1.

   Consider the following examples of subgroups.

   1. In the order-4 group C_4 (rotations by $\{0°, 90°, 180°, 270°\}$) the
elements $\{e,b\}$ (rotations by $\{0°, 180°\}$) form an order-2 subgroup C_2. This
is the only subgroup of C_4.

   2. The order-4 group D_2 includes three order-2 subgroups: $\{e,a\}$, $\{e,b\}$,
$\{e,c\}$. According to the above theorem the order-4 groups cannot have
order-3 subgroups, so we have here all their nontrivial proper subgroups.

3.   An order-6 group can have nontrivial proper subgroups of orders 2 and 3 only. In fact, the non-Abelian order-6 group D_3 includes four such subgroups, one of order 3 and three of order 2: $\{e,a,b\}$, $\{e,c\}$, $\{e,d\}$, $\{e,f\}$.

4.   The group of nonzero real numbers under multiplication includes various subgroups. Among these are the positive real numbers, the non-zero rational numbers, the positive rational numbers, and the integral powers of a fixed real number.

5.   The group of nonsingular complex $n \times n$ matrices under matrix multiplication has various ($n \times n$)-matrix subgroups. Among them are the unitary matrices, the nonsingular real matrices, the complex orthogonal matrices, the real orthogonal matrices, and the unimodular matrices.

Problems

1.   Prove that a necessary and sufficient condition for a subset H of a group G to be a subgroup of G is that for all elements h_1 and h_2 of H, $h_1 h_2^{-1}$ is an element of H.

2.   The *intersection* of sets H_1 and H_2, denoted $H_1 \cap H_2$, is the set of all elements each of which is a member of both H_1 and H_2. Prove that if H_1 and H_2 are subgroups of group G, then $H_1 \cap H_2$ is also a subgroup of G.

3.   Every group contains at least one element that commutes with all elements of the group, i.e., an element a such that $ag = ga$ for all elements g of the group. The identity e is such an element, and there may be more. The set of all such elements of a group is called the *center* of the group. Prove that the center of a group is a subgroup. What is the center of an Abelian group?

4.   An Abelian group cannot have a non-Abelian subgroup, but a non-Abelian group can have an Abelian subgroup. Find examples of the latter.

5.   The inclusion relation between groups is not an equivalence relation. But which of the properties of an equivalence relation does it fulfill?

6.   The group table of the order-8 group denoted D_4 is shown in Fig. 1.30. Find and identify all proper subgroups of D_4.

7.   The group table of the order-8 group denoted Q_4 (and called the *quaternion group*) is shown in Fig. 1.31. Find and identify all proper subgroups of Q_4.

8.   (a) Prove that the group of integers under addition is isomorphic with the group of even integers under addition. (b) Prove that the

e	a	b	c	d	f	g	h
a	b	c	e	f	g	h	d
b	c	e	a	g	h	d	f
c	e	a	b	h	d	f	g
d	h	g	f	e	c	b	a
f	d	h	g	a	e	c	b
g	f	d	h	b	a	e	c
h	g	f	d	c	b	a	e

Figure 1.30   Group table of D_4.

latter is a proper subgroup of the former. (c) If a group is isomorphic with a proper subgroup of itself, what can be stated about the order of the group?

9. Prove that each of the following sets of numbers forms a subgroup of the group of nonzero real numbers under multiplication: (a) positive real numbers; (b) nonzero rational numbers; (c) positive rational numbers; (d) the integral powers of a fixed real number.

10. Check if each of the following sets is a subgroup of the group of all real numbers under addition, the group of all nonzero real numbers under multiplication, or neither: (a) all integral multiples of a fixed real number; (b) $\{a^b\}$, where a is a fixed positive real number and b runs over the preceding set; (c) all nonzero pure imaginary complex numbers; (d) all irrational numbers together with the number 1; (e) all irrational numbers together with the number 0.

11. Prove that each of the following sets of $n \times n$ matrices is a subgroup of the group of nonsingular complex $n \times n$ matrices under matrix multiplication: (a) complex orthogonal matrices; (b) nonsingular di-

e	a	b	c	d	f	g	h
a	b	c	e	f	g	h	d
b	c	e	a	g	h	d	f
c	e	a	b	h	d	f	g
d	h	g	f	b	a	e	c
f	d	h	g	c	b	a	e
g	f	d	h	e	c	b	a
h	g	f	d	a	e	c	b

Figure 1.31   Group table of Q_4.

agonal matrices; (c) nonsingular real matrices; (d) nonzero complex scalar matrices. Are any of these subgroups of others of these?

12. Prove that the following groups of $n \times n$ matrices under matrix multiplication form an inclusion chain, where each group is a proper subgroup of each of the preceding groups: (a) nonsingular complex matrices; (b) unitary matrices; (c) real orthogonal matrices; (d) unimodular real orthogonal matrices; (e) diagonal unimodular real orthogonal matrices.

13. Is each of the following sets a subgroup of the group of all polynomials in one variable under addition? (a) All polynomials of degree less than 2; (b) all polynomials of even degree; (c) all polynomials containing only even powers of the variable; (d) all polynomials of odd degree; (e) all polynomials containing only odd powers of the variable; (f) all polynomials with rational coefficients.

Group Theory Continued

2.1. CONJUGACY, EQUIVALENCE CLASS, INVARIANT SUBGROUP, KERNEL

If for some pair of elements a and b of group G there exists a (not necessarily unique) element u of G such that

$$u^{-1}au = b,$$

then a and b are called *conjugate elements* in G. This conjugacy is denoted

$$a \equiv b.$$

It is an equivalence relation among group elements. (1) Conjugacy is reflexive, since every element a is conjugate with itself, $a \equiv a$ for all a in G. This is due to the fact that $e^{-1}ae = a$. (2) Conjugacy is symmetric, since for all a and b in G if $a \equiv b$, then $b \equiv a$, because if there exists an element u of G such that $u^{-1}au = b$, then $v^{-1}bv = a$ with $v = u^{-1}$. (3) And finally, conjugacy is transitive, since for all a, b, c in G if $a \equiv b$ and $b \equiv c$, then $a \equiv c$. If there exist elements u and v of G such that $u^{-1}au = b$ and $v^{-1}bv = c$, then $w^{-1}aw = c$ with $w = uv$.

Note that more than two elements might be conjugate with each other. The identity element cannot be conjugate with any other element. In an Abelian group there are no conjugate elements.

As an example, we find the conjugate elements of the order-6 group D_3 (Fig. 1.9). First we find all inverses: $a^{-1} = b$, $b^{-1} = a$, $c^{-1} = c$, $d^{-1} = d$, $f^{-1} = f$. We now conjugate each element of the group by all the others to find conjugates, for example:

$$c^{-1}ac = cac = b,$$
$$d^{-1}cd = dcd = f,$$
$$f^{-1}cf = fcf = d.$$

In this way we finally obtain the breakdown into conjugate elements:

$$e, a \equiv b, c \equiv d \equiv f.$$

A subset of elements of a group that consists of a complete set of mutually conjugate elements is called a *conjugacy class* of the group. The identity element, since it is not conjugate with any other element, is always in a conjugacy class of its own. In an Abelian group every element is in a conjugacy class of its own.

For example, the conjugacy classes of D_3 are $\{e\}$, $\{a,b\}$, $\{c,d,f\}$, as we just found.

At this point we introduce the concept of *equivalence class*, of which conjugacy class is a special case. If we have a set of elements for which an equivalence relation is defined, any subset that contains a complete set of mutually equivalent elements is called an equivalence class. It is easily shown that two different equivalence classes cannot have common elements. Thus an equivalence relation brings about a decomposition of the set for which it is defined, such that every element of the set is a member of one and only one equivalence class.

Two examples of equivalence class already noted are conjugacy classes, where the equivalence relation is conjugacy between group elements, and classes of groups with the same structure, where the equivalence relation is isomorphism of groups in the set of all groups.

Conversely, any decomposition of a set into subsets such that every element of the set is contained in one and only one subset defines an equivalence relation. The subsets may be declared, by fiat, equivalence classes, and the corresponding equivalence relation is simply that two elements are equivalent if and only if they belong to the same subset.

Returning now to conjugation, conjugation by a single element has an interesting and useful property. Let a, a', b, b', c, c' be elements of group G such that $a \equiv a'$, $b \equiv b'$, $c \equiv c'$ with respect to the same element u of G:

$$u^{-1} \begin{pmatrix} a \\ b \\ c \end{pmatrix} u = \begin{pmatrix} a' \\ b' \\ c' \end{pmatrix}.$$

Now assume that $ab = c$. Compose each side of this relation with u^{-1} on the left and with u on the right. The right-hand side becomes $u^{-1}cu = c'$. With some clearly indicated manipulations the left-hand side becomes

$$u^{-1}abu = u^{-1}aebu$$
$$= u^{-1}a(uu^{-1})bu$$
$$= (u^{-1}au)(u^{-1}bu)$$
$$= a'b'.$$

Thus $a'b' = c'$, and intragroup relations are invariant under conjugation by the same element. Note the analogy with invariance of algebraic relations among matrices under similarity transformations.

For example, in group D_3 $bd = f$. Under conjugation by c

$$c^{-1}\begin{pmatrix} b \\ d \\ f \end{pmatrix} c = c\begin{pmatrix} b \\ d \\ f \end{pmatrix} c = \begin{pmatrix} a \\ f \\ d \end{pmatrix},$$

and indeed $af = d$.

If H is a subset (not necessarily a subgroup) of group G, $g^{-1}Hg$ for some g in G denotes the subset of G consisting of elements $g^{-1}hg$, where h runs over all elements of H, i.e., the subset conjugate with H by g. If G is Abelian, $g^{-1}Hg = H$ for all g in G.

If H is a subgroup of G, then $g^{-1}Hg$ for any g in G is also a subgroup of G. (1) Since H is closed under the composition of G, so is $g^{-1}Hg$, since, as shown above, intragroup relations are invariant under conjugation by the same element (here g). (2) Associativity holds, since it holds in G. (3) $g^{-1}Hg$ contains the identity, since H contains it and $g^{-1}eg = e$. (4) Any element of $g^{-1}Hg$ is expressible as $g^{-1}hg$, where h is some element of H. Its inverse is $(g^{-1}hg)^{-1} = g^{-1}h^{-1}g$, which is an element of $g^{-1}Hg$, since h^{-1} is an element of H. Thus $g^{-1}Hg$ contains the inverses of all its elements. So $g^{-1}Hg$ is indeed a group if H is. And, again since intragroup relations are invariant under conjugation by the same element, conjugate subgroups are isomorphic.

For example, D_3 contains the subgroup $\{e,c\}$. The conjugate subgroup by a is

$$a^{-1}\{e,c\}a = b\{e,c\}a = \{e,f\}.$$

Both subgroups are isomorphic, since they are both of order 2, and there is only one abstract order-2 group, C_2.

If H is a subgroup of G and $g^{-1}Hg = H$ for all g in G, i.e., if H is conjugate only with itself, H is called an *invariant subgroup* (also *normal subgroup*).

As an example, the subgroup $\{e,a,b\}$ of D_3 is an invariant subgroup. Earlier in this section we found that elements a and b are conjugate with each other and with no other element, while the identity e is conjugate only with itself. Thus

$$g^{-1}\{e,a,b\}g = \{e,a,b\}$$

for all elements g of D_3 and the subgroup is proved invariant.

Note that the trivial subgroups $\{e\}$ and G of any group G are invariant. In an Abelian group all subgroups are invariant.

Now consider the homomorphism $G \to G'$. That subset of elements of G which is mapped to the identity element e' of G' is called the *kernel* of the homomorphism, as we mentioned in Section 1.5. A kernel is usually denoted K. Thus K consists of all g in G for which $g \to e'$. For any homomorphism $G \to G'$ the kernel K is an invariant subgroup of G, as we immediately prove.

First we prove that K is a subgroup of G.

1. Let a and b be any pair of elements of K, so that $a,b \to e'$. Denote the image of their composition $c\ (= ab)$ by c'. Preservation of structure requires

    $$\begin{array}{ccc} a & b & = c \\ \downarrow & \downarrow & \downarrow \\ e' & e' & = c' \end{array}$$

    But $e'e' = e'$, so $c' = e'$, $c \to e'$, and $c \in K$. Thus K is closed with respect to the composition of G.
2. Associativity holds in K, since it holds in G.
3. Is e an element of K? The identity e obeys $ea = a$ for all a in G. Let $e \to x'$ and $a \to a'$. Preservation of structure requires

    $$\begin{array}{ccc} e & a & = a \\ \downarrow & \downarrow & \downarrow \\ x' & a' & = a' \end{array}$$

    Thus $x'a' = a'$ for all a' in G', which means that $x' = e'$, since the identity is unique, so that $e \to e'$, and e is indeed an element of K.

4.   Let a be any element of K, so that $a \rightarrow e'$, and denote the image of its inverse by $(a^{-1})'$. By preservation of structure

$$
\begin{array}{ccc}
a & a^{-1} & = e \\
\downarrow & \downarrow & \downarrow \\
e' & (a^{-1})' & = e'
\end{array}
$$

Thus $(a^{-1})' = e'$, and a^{-1} is an element of K. So K contains the inverses of all its elements. That proves that K is a group.

We now prove that K is invariant. Let a be any element of K, so that $a \rightarrow e'$, and form the conjugation $g^{-1}ag = b$ for all elements g of G. Denote by b' the image of b, and denote by g' the image of g and by $(g^{-1})'$ the image of g^{-1}. Preservation of structure gives us $(g^{-1})' = g'^{-1}$ and

$$
\begin{array}{cccc}
g^{-1} & a & g & = b \\
\downarrow & \downarrow & \downarrow & \downarrow \\
g'^{-1} & e' & g' & = b'
\end{array}
$$

from which it follows that $b' = e'$ and $b \in K$. Thus $g^{-1}ag$ is an element of K for all a in K and all g in G, i.e., $g^{-1}Kg = K$ for all g in G, and K is invariant as claimed.

If the kernel of a homomorphism $G \rightarrow G'$ consists only of the identity element e, the mapping is, as will be seen in Section 2.4, one-to-one, and the homomorphism is actually an isomorphism. If the kernel is the group G itself, the homomorphism is the trivial one $g \rightarrow e'$ for all g in G, where G' is the order-1 group.

Consider some examples of homomorphism kernels as invariant subgroups.

1.   For the homomorphism $D_3 \rightarrow C_2$, presented as an example in Section 1.5, the kernel is $\{e, a, b\}$. It was shown in the preceding example in this section that this is an invariant subgroup.

2.   Consider the homomorphism, presented as an example in Section 1.5, of the group of nonsingular complex $n \times n$ matrices under matrix multiplication to the group of nonzero complex numbers under multiplication, where each matrix is mapped to the complex number that is the value of its determinant, $A \rightarrow |A|$. The kernel of this homomorphism is the subset of all unimodular $n \times n$ matrices. This subset is a subgroup, as mentioned in Section 1.6. Proof: (1) The product of unimodular $n \times n$ matrices is a unimodular $n \times n$ matrix $(|AB| = |A||B|)$, so the subset is closed under matrix multiplication. (2) Matrix multiplication is asso-

ciative. (3) The unit $n \times n$ matrix, the identity of matrix multiplication, is unimodular. (4) The inverse of any unimodular $n \times n$ matrix exists and is unimodular ($|A^{-1}| = 1/|A|$). So the kernel is a subgroup (which was proved in general in this section). Now check if it is an invariant subgroup (which it must be, as was proved in general). Let A be any nonsingular $n \times n$ matrix and U any unimodular $n \times n$ matrix. Form the conjugate element of U by A, $A^{-1}UA$ (this is just the similarity transformation of U by A; see the remark above). The value of its determinant is

$$|A^{-1}UA| = |A^{-1}||U||A|$$
$$= (1/|A|)|U||A|$$
$$= |U|$$
$$= 1$$

for all unimodular U and all nonsingular A, so $A^{-1}UA$ is always unimodular and an element of the kernel, which is thus confirmed to be an invariant subgroup.

3.   As seen in Section 1.5, the group of nonzero real numbers under multiplication is homomorphic to the group of numbers $\{1, -1\}$ under multiplication with mapping

$$\text{positive number} \rightarrow 1,$$

$$\text{negative number} \rightarrow -1.$$

The kernel is the subset of all positive real numbers, which is indeed a group under multiplication. It is trivially invariant, since multiplication of numbers is commutative.

Problems

1.   Prove the statement made in the text, that two different equivalence classes cannot have common elements, i.e., that two equivalence classes either are disjoint (have no element in common) or are one and the same.

2.   If H is an invariant subgroup of G and h is any element of H, prove that the conjugacy class of h in G is a subset of H.

3.   Prove that the center of a group is an invariant subgroup. (See Problem 3 of Section 1.6.)

4. For each of the equivalence relations of Problems 2 and 3 of Section 1.4 what are the equivalence classes?

5. For the order-8 group D_4 find all conjugate elements. Which of its proper subgroups are invariant? (See Problem 6 of Section 1.6.)

6. For the order-8 group Q_4 find all conjugate elements. Which of its proper subgroups are invariant? (See Problem 7 of Section 1.6.)

7. For the specific case of the homomorphism of the group of nonsingular complex $n \times n$ matrices to the group of unimodular complex $n \times n$ matrices show that the kernel is an invariant subgroup. (See Problem 4 of Section 1.5.)

2.2. COSET DECOMPOSITION

Let H be any proper subgroup of group G. Form the set aH, where a is any element of G that is not a member of H. This is the set of elements ah, where h runs over all the elements of H. Such a set aH is called a *left coset* of H in G. (Similarly, Ha is called a *right coset*.) Clearly, aH is not a subgroup of G, since it does not contain the identity element e. (For aH to contain e, H would have to contain a^{-1}, which it does not, since it is a group and does not contain a.) If H is of order m, then aH also contains m elements. Proof: aH certainly contains no more than m elements and could contain less only if $ah_1 = ah_2$ for some distinct pair of elements h_1 and h_2 of H. But $ah_1 = ah_2$ implies $h_1 = h_2$ by composition with a^{-1} on the left, which is a contradiction. Therefore, aH contains exactly m elements. In addition, no element of aH is also an element of H. Otherwise we would have $ah_1 = h_2$ for some pair of elements h_1 and h_2 of H, which implies $a = h_2 h_1^{-1}$ by composition with h_1^{-1} on the right. $a = h_2 h_1^{-1}$ means that a is a member of H, contrary to assumption. So H and aH have no element in common.

The *union* of a number of given sets is the set of all distinct elements of any of the given sets. The symbol $\cup$ is used to denote union. Form the union of H and aH, $H \cup aH$. The set $H \cup aH$ might or might not exhaust all the elements of G. If it does not, take any element b of G that is not a member of the set $H \cup aH$, i.e., b is an element neither of H nor of aH, and form the left coset bH. By reasoning similar to the above bH has exactly m elements and has no element in common with H or aH. Now form the union $H \cup aH \cup bH$. If this set does not exhaust G, continue forming left cosets in the same manner. If G is of finite order

n, we must eventually exhaust it by this procedure:

$$G = H \cup aH \cup bH \cup \cdots \cup kH.$$

Since H and each of its cosets have m elements per set and the sets have no element in common, we obtain the result that $n = ms$ for some integer s, as mentioned in Section 1.6. It is Lagrange's theorem (Joseph Louis Lagrange, French mathematician and astronomer, 1736–1813). (The whole proof could be performed just as well with right cosets.)

The decomposition of a group G into cosets of any subgroup will always be the same for left cosets as for right cosets if G is Abelian. If G is non-Abelian, the decompositions might be different, although not necessarily. However, the decomposition of G into left (or right) cosets of subgroup H is unique. By this we mean that, if we form a left coset lH of H with any element l of G that is not a member of H and is also not among the elements $a, b, \ldots, k$ used in the decomposition

$$G = H \cup aH \cup bH \cup \cdots \cup kH,$$

the coset lH will necessarily be one of the cosets $aH, bH, \ldots, kH$. To see this, note that l must be a member of one of these cosets, say aH. Since all elements of aH are of the form ah, where h runs over all elements of H, then $l = ah_1$ for some element h_1 of H. Thus $a = lh_1^{-1}$ (by composition with h_1^{-1} on the right), and all elements of aH are of the form

$$ah = (lh_1^{-1})h = l(h_1^{-1}h),$$

where h, and consequently also $h_1^{-1}h$, runs over all elements of H. Therefore $aH = lH$. So the same decomposition of G into left (or right) cosets of subgroup H is obtained whatever elements are used to form the cosets.

Examples of decomposition into cosets (refer to Section 1.1 for group tables of the groups involved) follow.

1. The order-4 group C_4 has a single nontrivial proper subgroup, $H = \{e,b\}$. Its left coset formed with a is

$$aH = \{ae,ab\} = \{a,c\}.$$

Its left coset formed with c is

$$cH = \{ce,cb\} = \{c,a\} = aH.$$

So the unique decomposition into cosets of H is

$$G = \{e,b\} \cup \{a,c\}.$$

Since this group is Abelian, decomposition into right cosets of H is the same, which is obvious in this example, since there is only a single possibility for any coset of H.

2. The order-4 group D_2 has three nontrivial proper subgroups: $\{e,a\}$, $\{e,b\}$, $\{e,c\}$. So we have three different decompositions into left cosets:

$$H = \{e,a\}, \, bH = cH = \{b,c\}, \, G = \{e,a\} \cup \{b,c\};$$

$$H = \{e,b\}, \, cH = aH = \{c,a\}, \, G = \{e,b\} \cup \{c,a\};$$

$$H = \{e,c\}, \, aH = bH = \{a,b\}, \, G = \{e,c\} \cup \{a,b\}.$$

Right and left cosets of the same subgroup are the same in this example, since D_2 is Abelian.

3. The non-Abelian order-6 group D_3 has the nontrivial proper subgroups $\{e,a,b\}$, $\{e,c\}$, $\{e,d\}$, $\{e,f\}$. So we have four decompositions into left cosets:

$$H = \{e,a,b\}, \, cH = \{c,d,f\}, \, G = \{e,a,b\} \cup \{c,d,f\};$$

$$H = \{e,c\}, \, aH = \{a,f\}, \, bH = \{b,d\}, \, G = \{e,c\} \cup \{a,f\} \cup \{b,d\};$$

$$H = \{e,d\}, \, aH = \{a,c\}, \, bH = \{b,f\}, \, G = \{e,d\} \cup \{a,c\} \cup \{b,f\};$$

$$H = \{e,f\}, \, aH = \{a,d\}, \, bH = \{b,c\}, \, G = \{e,f\} \cup \{a,d\} \cup \{b,c\}.$$

Since the group is non-Abelian, its decompositions into right cosets might be different from its decompositions into left cosets. Its four decompositions into right cosets are

$$H = \{e,a,b\}, \, Hc = \{c,f,d\}, \, G = \{e,a,b\} \cup \{c,f,d\};$$

$$H = \{e,c\}, \, Ha = \{a,d\}, \, Hb = \{b,f\}, \, G = \{e,c\} \cup \{a,d\} \cup \{b,f\};$$

$$H = \{e,d\}, \, Ha = \{a,f\}, \, Hb = \{b,c\}, \, G = \{e,d\} \cup \{a,f\} \cup \{b,c\};$$

$$H = \{e,f\}, \, Ha = \{a,c\}, \, Hb = \{b,d\}, \, G = \{e,f\} \cup \{a,c\} \cup \{b,d\}.$$

Thus decomposition into cosets of subgroup $\{e,a,b\}$ is the same for left and right, whereas decomposition into cosets of any of the order-2 subgroups is different for left and right.

Let H be an invariant proper subgroup of group G, so that $g^{-1}Hg = H$ for all g in G. By composition with g on the left, this relation can be put in the form $Hg = gH$ for all g in G, with the result that left and right cosets of any invariant subgroup are the same. (Be careful! $Hg = gH$ for all g in G does not mean that all elements of G commute with all elements of H, but rather that the set of elements hg, where h runs over all elements of H, is the same as the set of elements gh, where h runs over all elements of H, for all g in G.) Thus decomposition of a group into cosets of an invariant subgroup is the same whether left or right cosets are used. We saw this in the example of decomposition of D_3 into cosets of its invariant subgroup $\{e,a,b\}$. (In an Abelian group all subgroups are invariant, so left-coset decomposition and right-coset decomposition are the same for any subgroup.)

Problems

1. If H is a subgroup of G, prove that a necessary and sufficient condition for elements a and b of G to belong to the same right coset of H is $ab^{-1} \in H$ or, equivalently, $ba^{-1} \in H$. Find corresponding conditions for a and b to belong to the same left coset of H.

2. If H is a subgroup of G, we have seen that the left (or right) cosets of H are disjoint and exhaust G and are thus equivalence classes. Find a succinct formulation of the corresponding equivalence relation.

3. If H is a subgroup of G and

    $$G = H \cup aH \cup bH \cup \cdots \cup kH$$

    is a decomposition of G into left cosets of H, prove that

    $$G = H \cup Ha^{-1} \cup Hb^{-1} \cup \cdots \cup Hk^{-1}$$

    is a decomposition into right cosets.

4. Decompose the order-8 group D_4 into left and right cosets of each of its nontrivial proper subgroups. Point out the subgroups for which left-coset decomposition is the same as right-coset decomposition. Compare these with the invariant subgroups of D_4. (See Problem 5 of Section 2.1.)

5. The same as Problem 4 but for the order-8 group Q_4. (See Problem 6 of Section 2.1.)

6. Decompose the group of nonzero real numbers under multiplication

into cosets of each of the following subgroups: (a) positive real numbers; (b) $\{1, -1\}$.

7. Decompose the group of nonsingular complex $n \times n$ matrices under matrix multiplication into cosets of each of the following invariant subgroups: (a) the group of unimodular complex $n \times n$ matrices; (b) the kernel of its homomorphism to the group of unimodular complex $n \times n$ matrices. (See the second example and Problem 7 of Section 2.1.)

2.3. FACTOR GROUP

In Section 2.2 we found that decomposition of a group into left cosets of an invariant subgroup is the same as decomposition into right cosets. This fact has far-reaching consequences, as we will see in this section and the next.

We now define a composition of cosets of invariant subgroup H of group G. It is denoted $(aH)(bH)$ and is defined as the set of all elements $h'h''$, where h' runs over all elements of coset aH and h'' independently runs over all elements of coset bH. This can also be expressed as the set of elements $(ah_1)(bh_2) = ah_1bh_2$, where h_1 and h_2 independently run over all elements of H. Using associativity and equality of left and right cosets of H, we obtain the following string of equations for the composition of cosets aH and bH, as defined just now:

$$(aH)(bH) = a(Hb)H$$
$$= a(bH)H$$
$$= (ab)(HH)$$
$$= (ab)H.$$

The last equality follows from $HH = H$, since HH is the set of elements h_1h_2, where h_1 and h_2 independently run over all elements of H, which is just H itself, H being a subgroup. An expression such as $a(Hb)H$ is the set of elements $ah'h$, where h' runs over all elements of Hb and h independently runs over all elements of H, or, equivalently, the set of elements $a(h_1b)h_2 = ah_1bh_2$, where h_1 and h_2 independently run over all elements of H. After overcoming our unfamiliarity with the notation, the point of what we are doing here can be appreciated; the composition of coset aH and coset bH, $(aH)(bH)$, is also a coset of H, the coset $(ab)H$ (although some element other than ab might be used to form it in the decomposition of G). Thus the set of cosets of any invariant subgroup H of group G is closed under coset composition.

That being the case, let us see if coset composition has any additional interesting properties. Is it associative? Yes,

$$[(aH)(bH)](cH) = (aH)[(bH)(cH)]$$

as a direct result of associativity of the composition of G. The composition of any coset of H with H itself gives

$$H(aH) = (Ha)H$$
$$= (aH)H$$
$$= a(HH)$$
$$= aH$$

and

$$(aH)H = a(HH) = aH,$$

using associativity, equality of left and right cosets, and $HH = H$. So H has the characteristic property of an identity in coset composition; its composition with any coset in either order is just that coset. And what happens when we compose any coset aH with coset $a^{-1}H$?

$$(aH)(a^{-1}H) = a(Ha^{-1})H$$
$$= a(a^{-1}H)H$$
$$= (aa^{-1})(HH)$$
$$= eH$$
$$= H$$

and

$$(a^{-1}H)(aH) = a^{-1}(Ha)H$$
$$= a^{-1}(aH)H$$
$$= (a^{-1}a)(HH)$$
$$= eH$$
$$= H,$$

using associativity, equality of left and right cosets, properties of inverse and identity, and $HH = H$. Thus for any coset aH there exists a coset $a^{-1}H$ (although some element other than a^{-1} might be used to form it in the decomposition of G), such that their composition in either order is H. In other words, every coset has an inverse coset.

What do we have, if not a new group? We start with any group G and invariant proper subgroup H. We decompose G into cosets of H (left and right cosets being the same),

$$G = H \cup aH \cup bH \cup \ldots.$$

We consider the set consisting of H and its cosets, $\{H, aH, bH, \ldots\}$ (Is it clear that each element of this set is itself a set, i.e., this is a set of sets?), under coset composition. And, indeed, we found that we have (1) closure, (2) associativity, (3) existence of identity, (4) existence of inverses—in short, a group. This group is called the *factor group* (also *quotient group*) of G by H and is denoted G/H. (It is a good idea to review the development of the factor group and make sure it is clear how H's being a subgroup and the invariance of H each plays its role.) If G is of finite order n and H is of order m, the order of the factor group G/H is $s = n/m$ by Lagrange's theorem.

As an example we take for G the non-Abelian, order-6 group D_3 and its invariant subgroup $H = \{e,a,b\}$. The corresponding coset decomposition is

$$G = \{e,a,b\} \cup \{c,d,f\}.$$

The factor group G/H is the set of sets

$$G/H = \{\{e,a,b\}, \{c,d,f\}\}$$

under coset composition. It is an order-2 group, isomorphic with C_2.

Problem

Construct and identify (by finding an isomorphism with a more familiar group) the factor group G/H, where G and H are, respectively: (a) D_4 and each of its invariant subgroups; (b) Q_4 and each of its invariant subgroups; (c) the nonzero real numbers under multiplication and the positive real numbers; (d) the nonzero real numbers under multiplication and $\{1, -1\}$; (e) the nonsingular complex $n \times n$ matrices under matrix multiplication and the unimodular complex $n \times n$ matrices; (f) the nonsingular complex

$n \times n$ matrices under matrix multiplication and the kernel of the homomorphism of that group with the group of unimodular complex $n \times n$ matrices. (See Problems 4–7 of the preceding section.)

2.4. STRUCTURE OF HOMOMORPHISM

Let us return once more to homomorphism $G \to G'$ with kernel K. As was proved in Section 2.1, K is an invariant proper subgroup of G (excluding the trivial homomorphism $G \to C_1$). So we decompose G into cosets of K,

$$G = K \cup aK \cup bK \cup \ldots,$$

and construct the factor group

$$G/K = \{K, aK, bK, \ldots\}.$$

All elements of K are, by definition, mapped by the homomorphism to e', the identity element of G'. Consider the following structure-preservation diagram for any element a of G and all elements k of K, where a' is the image of a in G':

$$
\begin{array}{ccc}
a & k & = (ak) \\
\downarrow & \downarrow & \downarrow \\
a' & e' & = a'.
\end{array}
$$

Thus all members of coset aK are mapped to a', and all members of coset bK are mapped to b', the image of b in G', and so on.

Can members of different cosets of K be mapped to the same element of G'? No. To see this assume that b is not a member of coset aK, which is a necessary and sufficient condition for cosets aK and bK to be different. We then want to prove that, if $a \to a'$ and $b \to b'$, we must have $a' \neq b'$. So we assume the opposite, that $b \to a'$. By preservation of structure the image of b^{-1} in G' must be a'^{-1}. This gives us the structure-preservation diagram for the composition $b^{-1}a$:

$$
\begin{array}{ccc}
b^{-1} & a & = (b^{-1}a) \\
\downarrow & \downarrow & \downarrow \\
a'^{-1} & a' & = e' \quad .
\end{array}
$$

Thus $b^{-1}a$ is an element of K, say $b^{-1}a = k$. Composition with b on the left and k^{-1} on the right gives $ak^{-1} = b$. But if k is an element of K, so is k^{-1} (since K is a subgroup), so that b has the form ak_1, where k_1 is an element of K, and b is a member of coset aK in contradiction to our assumption. Thus whereas all members of the same coset of K are mapped by the homomorphism to the same element of G', members of different cosets have different images in G'. And since, by definition, all members of G' are images in the homomorphism, we thereby obtain a one-to-one correspondence between all cosets of K in G and all elements of G'. We also obtain the result that any homomorphism is an m-to-one mapping, where m is the order of the kernel.

This one-to-one mapping between the factor group G/K and G' is, in fact, an isomorphism. Preservation of structure is based on the relation

$$(aK)(bK) = (ab)K,$$

found in the preceding section, and is exhibited in the diagram

$$
\begin{array}{ccc}
(aK) & (bK) & = (ab)K \\
\downarrow & \downarrow & \downarrow \\
a' & b' & = (a'b')
\end{array}
$$

where a' and b' are the respective images of a and b in G'. The mapping looks like that of Fig. 2.1.

To summarize the result, given a homomorphism $G \rightarrow G'$ with kernel K, we have the isomorphism $G/K \sim G'$.

A converse result is obtained, if, instead of starting with a homomorphism, we start with group G and invariant proper subgroup H. Decompose G into cosets of H,

$$G = H \cup aH \cup bH \cup \ldots ,$$

$$
\begin{array}{ll}
\underline{G/K \sim G'} & \\
K \longleftrightarrow e' & \\
aK \longleftrightarrow a' & \\
bK \longleftrightarrow b' & \\
\bullet \quad \bullet & \\
\bullet \quad \bullet & \\
\bullet \quad \bullet &
\end{array}
$$

Figure 2.1. Isomorphism of G/K with G', where K is kernel of homomorphism of G to G'.

and construct the factor group

$$G/H = \{H, aH, bH, \ldots\}.$$

Consider the mapping from G to G/H, where each element of G is mapped to that coset of H of which it is a member. This is an m-to-one mapping, where m is the order of H, in which every element of G has an image in G/H and every element of G/H serves as an image. This mapping is a homomorphism. To see that structure is preserved note that the cosets of H to which elements a, b, ab of G belong can be represented as aH, bH, $(ab)H$, respectively (although different elements might be used to form them). Then, by the relation

$$(aH)(bH) = (ab)H,$$

proved in the preceding section, we have the structure-preservation diagram

$$
\begin{array}{ccc}
a & b & = (ab) \\
\downarrow & \downarrow & \downarrow \\
(aH) & (bH) & = (ab)H.
\end{array}
$$

The kernel of this homomorphism is H, since all elements of H, and only those, are mapped to H, the identity element of G/H.

To sum up this section, given a homomorphism $G \rightarrow G'$ with kernel K, we have the isomorphism $G/K \sim G'$. Conversely, given group G and proper invariant subgroup H, we have the homomorphism $G \rightarrow G/H$ with kernel H.

This should help clarify what goes on in a homomorphism and how homomorphism, kernel, invariant proper subgroup, and factor group are intimately interrelated.

Some examples follow to help clarify the clarification.

1. We again return to the non-Abelian, order-6 group D_3, discussed so often above, as G. Its only invariant proper subgroup is $H = \{e,a,b\}$, C_3. The factor group G/H, here D_3/C_3, is the order-2 group

$$G/H = \{H,cH\}$$

$$= \{\{e,a,b\}, \{c,d,f\}\},$$

isomorphic with the abstract order-2 group $G' = \{e',a'\}$, C_2. We then have the three-to-one homomorphism $G \rightarrow G/H$ with kernel H, as well, of course, as the three-to-one homomorphism $G \rightarrow G'$ with kernel H. The whole situation can be represented as in Fig. 2.2.

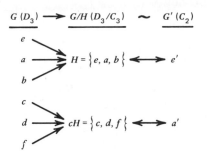

Figure 2.2. Homomorphism of G (D_3) to G/H (D_3/C_3), isomorphism of G/H (D_3/C_3) with G' (C_2), and homomorphism of G (D_3) to G' (C_2).

2. Consider the order-4 group C_4 as G, or its realization by the group of rotations about a common axis by $\{0°, 90°, 180°, 270°\}$. Since it is Abelian, its only proper subgroup $H = \{e, b\}$, C_2, or its realization by the subgroup of rotations by $\{0°, 180°\}$, is invariant. The factor group, C_4/C_2, is

$$G/H = \{H, aH\} = \{\{e, b\}, \{a, c\}\},$$

or its rotational realization

$$\{\{0°, 180°\}, \{90°, 270°\}\}.$$

It is isomorphic with the abstract group C_2, $G' = \{e', a'\}$, or its realization by rotations by $\{0°, 180°\}$. We then have the two-to-one homomorphism $G \to G/H$ with kernel H and the two-to-one homomorphism $G \to G'$ with kernel H, among the abstract groups as well as among their rotational realizations. All this can be expressed by the diagrams of Fig. 2.3.

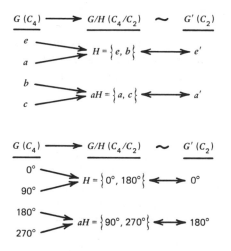

Figure 2.3. Homomorphism of G (C_4) to G/H (C_4/C_2), isomorphism of G/H (C_4/C_2) with G (C_2), and homomorphism of G (C_4) to G' (C_2): abstract groups and rotational realizations.

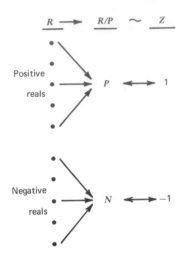

Figure 2.4. Homomorphism of multiplicative group of nonzero real numbers, R, to R/P, where P is multiplicative group of positive numbers, isomorphism of R/P with multiplicative group $Z = \{1, -1\}$, and homomorphism of R to Z. N is set of negative numbers.

3. We take an infinite-order example. The group of nonzero real numbers under multiplication, R, includes the group of positive numbers under multiplication, P, as a subgroup. P is an invariant subgroup, since R is Abelian. The corresponding coset decomposition of R is $R = P \cup a \times P$, where a is any number not in P, i.e., any negative number. The coset $a \times P$ is just the set of negative numbers, N. The order-2 factor group is $R/P = \{P, N\}$. It is isomorphic with any other order-2 group, C_2, say $Z = \{1, -1\}$ under multiplication. We then have the infinity-to-one homomorphisms $R \to R/P$ and $R \to Z$ with P as the kernel. Refer to Fig. 2.4.

Problem

In light of the discussion of this section analyze the relations among G, H, and G/H for each part of the problem of Section 2.3 except part (c), which was analyzed in the third example of the present section.

2.5. GENERATORS

Any subset of elements of a finite-order group such that they, their positive powers, and their compositions (and repeated compositions, as many times as necessary) produce the whole group is called a *set of generators* of the group. For a given group a set of generators is not unique. For example, the set of all elements of a group is a set of generators, and so

is the set of all elements excluding the identity. However, there are sometimes smaller sets, and there are always minimal sets. Minimal sets of
generators are not in general unique either. If a minimal set of generators
of a group is a single element, so that this element and its positive powers
produce the whole group, such a group is called a *cyclic group,* and this
element is called a *generating element* of the group. Since all elements
of a cyclic group are powers of a single element, a cyclic group is always
Abelian. It can be shown that a group of prime order must be cyclic. Thus
there is only a single abstract group of any prime order, the cyclic group
of that order, which is Abelian. Finite-order groups of rotations about a
common axis are cyclic. A possible generating element is the smallest
(nonzero) rotation of each group. Cyclic groups are denoted C_n, where
n is the order. Thus C_n can be realized by the group of rotations about
a common axis by $\{0°, 360°/n, 2 \times 360°/n, \ldots, (n - 1) \times 360°/n\}$. (The
notation D_n is for the *dihedral groups,* of order $2n$. We do not discuss
these here but present their general definition in Chapter 3.)

Examples of minimal sets of generators and cyclic groups include the
following.

1. The order-2 group C_2 is cyclic, since $a^2 = e$, so a is its generating
element. For its realization by $\{1, -1\}$ under multiplication the number
-1 is the generating element, since $(-1) \times (-1) = 1$. In terms of its
realization by rotations about a common axis by $\{0°, 180°\}$, rotation by
$180°$ is the generating element, since two consecutive rotations by $180°$
produce a rotation by $360°$, the identity transformation.

2. The order-3 group C_3 is cyclic, since $a^2 = b$ and $a^3 = e$, or b^2
$= a$ and $b^3 = e$, so that a and b are both generating elements. In terms
of its realization by rotations by $\{0°, 120°, 240°\}$ about a common axis,
rotation by $120°$ is a generating element, since two consecutive rotations
by $120°$ produce a rotation by $240°$, and three consecutive rotations result
in a rotation by $360°$, which is rotation by $0°$. Rotation by $240°$ is also a
generating element. Two consecutive rotations by $240°$ produce rotation
by $480°$, which is rotation by $120°$, and three consecutive $240°$ rotations
make a rotation by $720° = 2 \times 360°$, which is the identity. Note that both
2 and 3 are primes and that there is only a single group of each order and
it is cyclic.

3. The order-4 group C_4 is cyclic, since $a^2 = b$, $a^3 = c$, and $a^4 =$
e, or $c^2 = b$, $c^3 = a$, and $c^4 = e$. Thus a and c are generating elements.
However, b is not a generating element, since $b^2 = e$ and higher powers
of b will not produce a and c. In terms of this group's realization by
rotations about a common axis by $\{0°, 90°, 180°, 270°\}$, the generating
elements are rotation by $90°$ and rotation by $270°$. Rotation by $180°$ is not

a generating element, since two consecutive rotations by 180° produce the identity, and the other rotations are not produced by consecutive rotations by 180°.

4. Since 4 is not a prime, there may be noncyclic groups of this order, and indeed there is one, D_2. Minimal sets of generators of D_2 are $\{a,b\}$ ($a^2 = e$, $ab = c$), $\{a,c\}$ ($a^2 = e$, $ac = b$), and $\{b,c\}$ ($b^2 = e$, $bc = a$). In terms of its realization by reflections and rotation, presented in Section 1.1, a minimal set of generators consists of two reflections or the rotation and either reflection.

5. The order-5 group C_5 has prime order and is cyclic. Any one of its nonidentity elements is a generating element, for example, c, $c^2 = a$, $c^3 = d$, $c^4 = b$, $c^5 = e$. In the rotational realization of this group any nonzero rotation is a generating element.

6. The order-6 group D_3 is non-Abelian and therefore not cyclic. Two of its minimal sets of generators are $\{a,c\}$ ($a^2 = b$, $a^3 = e$, $ac = f$, $a^2c = d$) and $\{d,f\}$ ($d^2 = e$, $df = a$, $fd = b$, $dfd = c$).

Problems

1. Prove that a group of prime order must be cyclic.

2. Close your eyes and find all abstract order-7 groups. Display their group tables.

3. Construct the group table of the cyclic group of order 8, C_8.

4. Find at least two minimal sets of generators for the order-8 group D_4 and construct all the group elements from each set. (See Problem 6 of Section 1.6.)

5. $\{a,b\}$ is a minimal set of generators of a certain group, where $a^4 = b^2 = e$ and $ba = a^3b$. Construct the composition table for the set $\{e,a,a^2,a^3,b,ab,a^2b,a^3b\}$ and show that the set is closed. Since the set contains $\{a,b\}$ and is closed, it is the group under consideration. Show that this is D_4. (See Problem 6 of Section 1.6.)

6. The same as Problem 5, except that $a^4 = e$, $a^2 = b^2$, $ba = a^3b$, and the group is Q_4. (See Problem 7 of Section 1.6.)

2.6. DIRECT PRODUCT

The *direct product* of two groups G and G' is denoted $G \times G'$ and is the group consisting of elements that are ordered pairs (g,g'), where g and

g' run independently over all elements of G and G', respectively. If the orders of G and G' are n and n', respectively, the order of $G \times G'$ is nn'. Composition in $G \times G'$ is defined

$$(a,a')(b,b') = (ab,a'b')$$

for all a, b in G and all a', b' in G'. This assures (1) closure and (2) associativity. (3) The identity of $G \times G'$ is (e,e'), where e and e' are the respective identities of G and G'. (4) The inverse of any element (g,g') of $G \times G'$ is (g^{-1},g'^{-1}). Thus $G \times G'$ is proved to be a group as claimed. Among the possible proper subgroups of $G \times G'$ there is one isomorphic with G, consisting of elements of the form (g,e') for all g in G, and another isomorphic with G', consisting of elements of the form (e,g') for all g' in G'. Both are invariant. Clearly the direct-product group $G' \times G$ is isomorphic with $G \times G'$. Higher-degree direct-product groups are formed similarly. For instance, $G \times G' \times G''$ is the group of ordered triples (g,g',g''), where g, g', g'' run independently over all elements of G, G', G'', respectively. Composition is

$$(a,a',a'')(b,b',b'') = (ab,a'b',a''b'').$$

Consider the following examples.

1.   The direct product $C_2 \times C_2$ is an order-4 group. But which? The elements of this group are (e,e), (e,a), (a,e), and (a,a), with $a^2 = e$. Figure 2.5 is the group table. The total domination of the group-table diagonal by the identity element discloses that this is the structure of D_2. One possible isomorphism mapping is

$$(e,e),\ (e,a),\ (a,e),\ (a,a)$$
$$\updownarrow \quad \updownarrow \quad \updownarrow \quad \updownarrow$$
$$e',\quad a',\quad b',\quad c'\ .$$

2.   The direct product $C_2 \times C_3$ is an order-6 group. Since the constituent groups are Abelian, so is their direct product. Therefore, this is

(e,e)	(e,a)	(a,e)	(a,a)
(e,a)	(e,e)	(a,a)	(a,e)
(a,e)	(a,a)	(e,e)	(e,a)
(a,a)	(a,e)	(e,a)	(e,e)

Figure 2.5.   Group table of $C_2 \times C_2$.

(e,e')	(a,a')	(e,b')	(a,e')	(e,a')	(a,b')
(a,a')	(e,b')	(a,e')	(e,a')	(a,b')	(e,e')
(e,b')	(a,e')	(e,a')	(a,b')	(e,e')	(a,a')
(a,e')	(e,a')	(a,b')	(e,e')	(a,a')	(e,b')
(e,a')	(a,b')	(e,e')	(a,a')	(e,b')	(a,e')
(a,b')	(e,e')	(a,a')	(e,b')	(a,e')	(e,a')

Figure 2.6. Group table of $C_2 \times C_3$.

not the order-6 group D_3, discussed so frequently already. Its elements
are (e,e'), (a,a'), (e,b'), (a,e'), (e,a'), (a,b'), where $\{e,a\}$ forms the order-
2 group and $\{e',a',b'\}$ makes up the order-3 group. The group table is seen
in Fig. 2.6. If we relabel the elements e'', a'', b'', c'', d'', f'', respectively,
the group tables takes the form of Fig. 2.7. This is clearly the cyclic group
of order 6, C_6. Its generating elements are a'' and f'', or in $C_2 \times C_3$
notation (a,a') and (a,b'). C_6 can be realized by the group of rotations
about a common axis by $\{0°, 60°, 120°, 180°, 240°, 300°\}$, whose generating
elements are rotation by $60°$ and rotation by $300°$. C_6 and D_3 are the only
abstract groups of order 6.

3.   The group of complex numbers under addition is isomorphic with
the direct product of the group of real numbers under addition with itself.

4.   The group of 3-vectors under vector addition is isomorphic with
the direct product of the group of real numbers under addition with itself
twice.

Problems

1.   In the direct-product group $G \times G'$ prove that the two subgroups
consisting of elements of the form (g,e') and (e,g'), isomorphic with
G and G', respectively, are both invariant. Decompose $G \times G'$ into
cosets of each. Identify the factor groups $G \times G'/G$ and $G \times G'/G'$.

e''	a''	b''	c''	d''	f''
a''	b''	c''	d''	f''	e''
b''	c''	d''	f''	e''	a''
c''	d''	f''	e''	a''	b''
d''	f''	e''	a''	b''	c''
f''	e''	a''	b''	c''	d''

Figure 2.7. Group table of C_6.

2.   Construct the group table of the order-8 group $C_2 \times C_4$. Find a minimal set of generators and reconstruct the group table in terms of it. Prove that this group is not isomorphic with C_8, D_4, or Q_4.

3.   Construct the group table of the order-8 group $C_2 \times C_2 \times C_2$. Find a minimal set of generators and reconstruct the group table in terms of it. Prove that this group is not isomorphic with C_8, $C_2 \times C_4$, D_4, or Q_4.

4.   We have now collected five order-8 groups: C_8, $C_2 \times C_4$, $C_2 \times C_2 \times C_2$, D_4, and Q_4. These are, in fact, the only abstract groups of order 8. So the order-8 group $C_2 \times D_2$ must be isomorphic with one of them. With which?

2.7.   PERMUTATIONS, SYMMETRIC GROUPS

A *permutation* is a rearrangement of objects. Imagine, for example, that we have four objects in positions 1, 2, 3, 4, and imagine that we permute (i.e., rearrange) them so that the object in position 1 is placed in position 4, the object in position 2 is placed in position 1, the object in position 3 is left where it is, and the object in position 4 is placed in position 2. Figure 2.8 should help. Such a permutation is denoted $\begin{pmatrix} 1234 \\ 4132 \end{pmatrix}$, where the top row lists the positions, and each number in the bottom row is the final location of the object originally located at the position whose number is directly above it in the top row. The top row can be in any order, so both $\begin{pmatrix} 2431 \\ 1234 \end{pmatrix}$ and $\begin{pmatrix} 3142 \\ 3421 \end{pmatrix}$, for instance, also denote the permutation in our example. A general permutation of n objects can be denoted

$$\begin{pmatrix} 1 \; 2 \; \dots \; n \\ p_1 p_2 \dots p_n \end{pmatrix},$$

where $p_1, p_2, \dots, p_n$ are the numbers 1, 2, $\dots$, n in some order (i.e., some permutation of 1, 2, $\dots$, n), which directs us to move the object in position 1 to position p_1, the object in position 2 to position p_2, and

Figure 2.8.   Permutation $\begin{pmatrix} 1234 \\ 4132 \end{pmatrix}$.

so on. The top row can be in any order, so long as p_i appears under i for $i = 1, 2, \ldots, n$.

Permutations of the same number of objects are composed by consecutive application; the second permutation acts on the result of the first. The composition of permutations $a = \begin{pmatrix} 1234 \\ 4132 \end{pmatrix}$ and $b = \begin{pmatrix} 1234 \\ 1342 \end{pmatrix}$, for example, with permutation a acting first and permutation b acting on the result of a, is denoted

$$ba = \begin{pmatrix} 1234 \\ 1342 \end{pmatrix}\begin{pmatrix} 1234 \\ 4132 \end{pmatrix}.$$

(The order of application of permutations, as well as of any other kind of transformation, is specified from right to left. This is discussed in Chapter 3.) The composition is the permutation obtained by following the itinerary of each object through the permutations to which it is subjected. In the example the object originally in position 1 is carried to position 4 by permutation a and from there to position 2 by b. The object originally in position 2 is shifted to position 1 by a and then left in position 1 by b. And so on to obtain $ba = \begin{pmatrix} 1234 \\ 2143 \end{pmatrix}$. A useful trick is to reorder the rows of the second permutation to make the top row match the bottom row of the first permutation. The combined permutation is then obtained by "canceling" the bottom row of the first permutation with the top row of the second, leaving the top row of the first permutation and the bottom row of the second as the result. In the example

$$ba = \begin{pmatrix} 1234 \\ 1342 \end{pmatrix}\begin{pmatrix} 1234 \\ 4132 \end{pmatrix}$$

$$= \begin{pmatrix} 4132 \\ 2143 \end{pmatrix}\begin{pmatrix} 1234 \\ 4132 \end{pmatrix}$$

$$= \begin{pmatrix} 1234 \\ 2143 \end{pmatrix}.$$

Also

$$ab = \begin{pmatrix} 1234 \\ 4132 \end{pmatrix}\begin{pmatrix} 1234 \\ 1342 \end{pmatrix}$$

$$= \begin{pmatrix} 1342 \\ 4321 \end{pmatrix}\begin{pmatrix} 1234 \\ 1342 \end{pmatrix}$$

$$= \begin{pmatrix} 1234 \\ 4321 \end{pmatrix}.$$

For general permutations of n objects

$$a = \begin{pmatrix} 1 & 2 & \ldots & n \\ p_1 p_2 & \ldots & p_n \end{pmatrix} \quad \text{and} \quad b = \begin{pmatrix} 1 & 2 & \ldots & n \\ q_1 q_2 & \ldots & q_n \end{pmatrix}$$

the composition is

$$
\begin{aligned}
ba &= \begin{pmatrix} 1 & 2 & \ldots & n \\ q_1 q_2 & \ldots & q_n \end{pmatrix} \begin{pmatrix} 1 & 2 & \ldots & n \\ p_1 p_2 & \ldots & p_n \end{pmatrix} \\
&= \begin{pmatrix} p_1 & p_2 & \ldots & p_n \\ q_{p_1} q_{p_2} & \ldots & q_{p_n} \end{pmatrix} \begin{pmatrix} 1 & 2 & \ldots & n \\ p_1 p_2 & \ldots & p_n \end{pmatrix} \\
&= \begin{pmatrix} 1 & 2 & \ldots & n \\ q_{p_1} q_{p_2} & \ldots & q_{p_n} \end{pmatrix} .
\end{aligned}
$$

Similarly,

$$ab = \begin{pmatrix} 1 & 2 & \ldots & n \\ p_{q_1} p_{q_2} & \ldots & p_{q_n} \end{pmatrix} .$$

Thus the composition of permutations of n objects is also a permutation of n objects. Composition of permutations is not, in general, commutative, as we see in the example or in the general case. It is, however, associative, as we could easily prove by composing three permutations and obtaining three tiers of indices in the bottom row, which we mercifully refrain from doing. (The composition of transformations of any kind by consecutive application is associative, as proved in Chapter 3.)

The "nonpermutation" $\begin{pmatrix} 1 2 \ldots n \\ 1 2 \ldots n \end{pmatrix}$ serves as the identity permutation of n objects. The inverse of permutation

$$\begin{pmatrix} 1 & 2 & \ldots & n \\ p_1 p_2 & \ldots & p_n \end{pmatrix}$$

is simply

$$\begin{pmatrix} p_1 p_2 & \ldots & p_n \\ 1 & 2 & \ldots & n \end{pmatrix} ;$$

each clearly undoes the other.

What could we be leading up to, if not that the set of all permutations of n objects, under composition of consecutive permutation, forms a group? It is called the *symmetric group* of degree n and is denoted S_n.

$$S_3 = \left\{ \begin{pmatrix} 123 \\ 123 \end{pmatrix}, \begin{pmatrix} 123 \\ 231 \end{pmatrix}, \begin{pmatrix} 123 \\ 312 \end{pmatrix}, \begin{pmatrix} 123 \\ 132 \end{pmatrix}, \begin{pmatrix} 123 \\ 321 \end{pmatrix}, \begin{pmatrix} 123 \\ 213 \end{pmatrix} \right\}$$
$$\updownarrow \quad\quad \updownarrow \quad\quad \updownarrow \quad\quad \updownarrow \quad\quad \updownarrow \quad\quad \updownarrow$$
$$D_3 = \{ \quad e, \quad\quad a, \quad\quad b, \quad\quad c, \quad\quad d, \quad\quad f, \quad \}$$

Figure 2.9. Isomorphism of S_3 with D_3.

In the preceding paragraphs we (1) proved closure, (2) claimed associativity, and proved (3) existence of identity and (4) existence of inverses. The order of S_n is $n!$. It is non-Abelian for $n > 2$.

It should be clear that $S_n \supset S_m$ for $n \geq m$.

Examples of symmetric groups include the following.

1. S_1 is the order-1 group of the identity permutation $\begin{pmatrix} 1 \\ 1 \end{pmatrix}$.

2. S_2 consists of the permutations $\begin{pmatrix} 12 \\ 12 \end{pmatrix}$ and $\begin{pmatrix} 12 \\ 21 \end{pmatrix}$ and is isomorphic with C_2. It is Abelian.

3. S_3 is an order-6 group. It is non-Abelian, so it must be isomorphic with D_3. (See the second example of the preceding section.) A mapping is given in Fig. 2.9. (See Fig. 1.9.)

4. S_4 is of order 24, too large to consider here, and higher-degree symmetric groups have even larger, and rapidly increasing, orders. For example, $11! = 39,916,800$, and $12!$ exceeds the capacity of an eight-digit calculator.

Problems

1. Prove what was declared in the text as being clear, that $S_n \supset S_m$ for $n \geq m$.

2. Given the permutations $a = \begin{pmatrix} 12345 \\ 53214 \end{pmatrix}$ and $b = \begin{pmatrix} 12345 \\ 32154 \end{pmatrix}$: (a) Calculate the permutations a^2, b^2, ab, and ba; (b) find the permutations a^{-1}, b^{-1}, $(ab)^{-1}$, and $(ba)^{-1}$, calculate $b^{-1}a^{-1}$ and $a^{-1}b^{-1}$, and compare; (c) solve for permutations c and d, where $ac = b = da$; (d) solve for permutations f and g, where $bf = a = gb$.

3. Check associativity of composition of permutations by comparing $(ab)c$ and $a(bc)$ for the special case of $a = \begin{pmatrix} 123456 \\ 415263 \end{pmatrix}$, $b = \begin{pmatrix} 123456 \\ 361542 \end{pmatrix}$, $c = \begin{pmatrix} 123456 \\ 146352 \end{pmatrix}$.

4.   Calculate all compositions of elements of S_3 and construct a group table for this group.

2.8.   CYCLES, TRANSPOSITIONS, ALTERNATING GROUPS

For another way of looking at permutations we consider the permutation

$$\begin{pmatrix} 12345678 \\ 45132768 \end{pmatrix},$$

an element of S_8. The object originally in position 1 is moved to position 4. The object originally in position 4 must be evicted and is placed in position 3. The object evicted from position 3 is moved into position 1, closing the circle. Such a chain of replacements is called a *cycle*. This cycle is $1 \to 4 \to 3 \to 1$. It is denoted (143). The other cycles of our permutation are $2 \to 5 \to 2$, denoted (25), $6 \to 7 \to 6$, denoted (67), and $8 \to 8$, denoted (8). A cycle is a permutation, a cyclic permutation. For example,

$$(143) = \begin{pmatrix} 12345678 \\ 42135678 \end{pmatrix}, \qquad (25) = \begin{pmatrix} 12345678 \\ 15342678 \end{pmatrix}, \qquad (8) = \begin{pmatrix} 12345678 \\ 12345678 \end{pmatrix}.$$

Since a cycle is a closed replacement chain, it may be written starting with any location in the chain. For example, $(143) = (431) = (314)$ and $(25) = (52)$. However, the cyclic order must be kept. For example, $(143) \neq (134)$.

Thus any permutation consists of a number of cycles. These cycles all commute with each other, since they apply to different locations, i.e., no two cycles of a permutation contain the same symbol. Any permutation can be denoted as a composition of its cycles, in any order since they commute. For our example

$$\begin{pmatrix} 12345678 \\ 45132768 \end{pmatrix} = (67)(143)(25)(8)$$

or just $(67)(143)(25)$, since a cycle of one is just the identity and can be dropped (as long as the degree of the permutation is remembered). Or, equally well,

$$\begin{pmatrix} 12345678 \\ 45132768 \end{pmatrix} = (431)(67)(52)$$

$$= (25)(314)(76), \text{ etc.}$$

The decomposition of a permutation into cycles clearly is unique (although it may be denoted in more than one way).

A cycle of two, an interchange of two objects, is called a *transposition*. Any cycle can be decomposed into transpositions. For example,

$$(143) = (13)(14) = (41)(43).$$

(Read transposition compositions from right to left!) In such decompositions the transpositions do not commute, since they have symbols in common, so their order is important. Also, as seen in the example, decomposition into transpositions is not unique. But an n-cycle always decomposes into a minimum of $n - 1$ transpositions. Here is a recipe for such a decomposition:

$$(12 \ldots n) = (1n) \ldots (13)(12).$$

And if an n-cycle is decomposed into more than $n - 1$ transpositions, it turns out that the number of transpositions can differ from $n - 1$ only by a multiple of 2. Thus an n-cycle can be decomposed only into an odd number of transpositions if n is even, or an even number of transpositions if n is odd.

In this way any permutation can be decomposed into a composition of transpositions; it is first decomposed into cycles, then all the cycles are decomposed into transpositions. For example,

$$\begin{pmatrix} 12345678 \\ 45132768 \end{pmatrix} = (67)(143)(25)$$
$$= (67)(13)(14)(25).$$

The decomposition is not unique, and neither is the number of transpositions in the decomposition. What is uniquely determined, however, is the minimal number of transpositions in the decomposition and the evenness or oddness of the number of transpositions in the decomposition. If the number is even or odd, the permutation is said to be, respectively, an even or odd permutation, or to possess, respectively, even or odd *parity*. [You might be familiar with a definition of the determinant of a square matrix in which the sign of each term is determined by the parity of the permutation of indices in that term through the function $(-1)^h$, where h is the number of transpositions involved in the permutation.]

The composition of two even or two odd permutations is an even permutation. The composition of an even and an odd permutation is an

$D_3 \sim S_3$		Parity	
$e \leftrightarrow \begin{pmatrix} 123 \\ 123 \end{pmatrix} = (1)$		Even	
$a \leftrightarrow \begin{pmatrix} 123 \\ 231 \end{pmatrix} = (123) = (13)\,(12)$		Even	$A_3 \sim C_3$
$b \leftrightarrow \begin{pmatrix} 123 \\ 312 \end{pmatrix} = (132) = (12)\,(13)$		Even	
$c \leftrightarrow \begin{pmatrix} 123 \\ 132 \end{pmatrix} = (23)$		Odd	
$d \leftrightarrow \begin{pmatrix} 123 \\ 321 \end{pmatrix} = (13)$		Odd	
$f \leftrightarrow \begin{pmatrix} 123 \\ 213 \end{pmatrix} = (12)$		Odd	

Figure 2.10. Decomposition of elements of S_3 into transpositions with parities indicated. Isomorphism of S_3 with D_3 is shown for convenience. Set of even-parity elements, A_3, is invariant subgroup of S_3 and is isomorphic with C_3, the invariant subgroup of D_3.

odd permutation. The identity permutation is even (zero transpositions). The inverse of an even or odd permutation is, respectively, even or odd. Thus the set of all even permutations of n objects forms a subgroup of S_n (the odd permutations do not), called the *alternating group* of degree n and denoted A_n. The order of A_n is $\frac{1}{2}n!$ A_n is, in fact, an invariant subgroup of S_n. The factor group S_n/A_n is isomorphic with C_2.

For example, in Fig. 2.10 we decompose the elements of S_3 into transpositions and indicate their parities. For convenience we also show the isomorphism mapping to D_3 (see Fig. 1.9). The invariant subgroup A_3 of S_3 is isomorphic with and mapped to the invariant subgroup $\{e,a,b\}$, C_3, of D_3.

We use this example to elucidate the significance of conjugacy among elements of S_n. For D_3 we found the conjugacy classes e, $a \equiv b$, $c \equiv d \equiv f$. In terms of its realization by S_3, the conjugacy classes are (1), (123) $\equiv$ (132), (23) $\equiv$ (13) $\equiv$ (12). Thus here and for general S_n, conjugate permutations have the same cycle structure. They differ only in the labeling of the symbols which are permuted. This is analogous with similarity transformations in linear algebra. There square, column, and row matrices related by a similarity transformation can be interpreted as representing the same object, an operator or a vector, but expressed with respect to different bases, i.e., different "names" or "labels" for the same objects.

Problems

1. Prove the statements made in the text: (a) The composition of two even or two odd permutations is an even permutation; (b) the composition of an even and an odd permutation is an odd permutation; (c) the parity of permutation a is the same as that of a^{-1}; (d) A_n is an *invariant* subgroup of S_n.

2. Referring to Problem 2 of Section 2.7, take the results obtained there and: (a) express a, b, a^2, b^2, ab, ba in cycle form, then recalculate the compositions in cycle representation; (b) express a, b, a^2, b^2, ab, ba in the form of transpositions, then recalculate the compositions in this representation. (c) Check the parities of a, b, a^2, b^2, ab, ba. (*Hints:* Remember that the order of operation of consecutive permutations is from right to left. When composing transpositions, the result is conveniently obtained in cycle form and then converted to transpositions. Remember that representation of a permutation in transposition form is not unique.)

2.9. CAYLEY'S THEOREM

Consider the group table of any group G of finite order n. For example, the group D_3, whose table is repeated in Fig. 2.11. If the first row is e, a, b, . . . , the row starting with g is obtained from the first row by composition with g on the left: $ge (= g)$, ga, gb, ... From this it follows directly that (1) each row contains all the elements of the first row, but in some other arrangement (unless $g = e$), and (2) rows starting with different elements are different. As for (1), if any row did not contain all the elements of the first row, we would have, say, $ga = gb$ for the element g at the left of that row and some pair of elements a, b. Composing both

e	a	b	c	d	f
a	b	e	f	c	d
b	e	a	d	f	c
c	d	f	e	a	b
d	f	c	b	e	a
f	c	d	a	b	e

Figure 2.11. Group table of D_3.

$$\begin{pmatrix} e & a & b & \cdots \\ g_2 & g_2a & g_2b & \cdots \end{pmatrix} \overset{g_2}{\leftrightarrow} \begin{pmatrix} e & a & b & \cdots \\ g_1 & g_1a & g_1b & \cdots \end{pmatrix} \overset{g_1}{\leftrightarrow}$$

$$= \overset{g_2}{\leftrightarrow} \begin{pmatrix} g_1 & g_1a & g_1b & \cdots \\ g_2g_1 & g_2(g_1a) & g_2(g_1b) & \cdots \end{pmatrix} \overset{g_1}{\leftrightarrow} \begin{pmatrix} e & a & b & \cdots \\ g_1 & g_1a & g_1b & \cdots \end{pmatrix}$$

$$=$$

$$= \overset{(g_2g_1)}{\leftrightarrow} \begin{pmatrix} e & a & b & \cdots \\ g_2g_1 & (g_2g_1)a & (g_2g_1)b & \cdots \end{pmatrix}$$

Figure 2.12. Structure-preservation diagram for Cayley's theorem.

sides of this equation with g^{-1} on the left, we would then have $a = b$, which is a contradiction, since the first row of the group table contains n different symbols for n distinct elements. The arrangement of the elements in the row starting with g is different from that of the first row, simply because the first row starts with e and the other does not (unless $g = e$, and we are talking about the first row anyway). The second result is trivial, since rows starting with different elements are different simply in that they start with different elements.

What we are getting at, however, is that to each element of a finite group we can assign a unique permutation of n objects in a one-to-one correspondence; to every element g we assign the permutation whose top row is the top row of the group table and whose bottom row is that row of the group table starting with g,

$$g \leftrightarrow \begin{pmatrix} e & a & b \ldots \\ g & ga & gb \ldots \end{pmatrix}.$$

(This is indeed a permutation, although we have become used to denoting permutations with numbers rather than with letters. To make it look more familiar one might replace the letters $e, a, b, \ldots$ by the numbers 1, 2, $\ldots, n$ according to some cipher.) From the discussion of the preceding paragraph it is clear that this is indeed a one-to-one mapping of all elements of G onto a subset of S_n. In fact, this mapping is an isomorphism, and this subset of S_n is therefore a subgroup of S_n. To see that structure is preserved consider Fig. 2.12, keeping in mind our trick for composing permutations. (The whole discussion could be carried out just as well with columns instead of rows of a group table.) Our result, known as Cayley's theorem (Arthur Cayley, British mathematician, 1821–1895), is that every group of finite order n is isomorphic with a subgroup of S_n. Cayley's theorem is helpful for finding groups of a given order.

Cayley's theorem also helps us elucidate the significance of conjugacy among group elements. We saw previously that conjugate permutations have the same cycle structure, i.e., they are, in a sense, the same permutation but differ only in the labeling of the permuted symbols. By Cayley's theorem conjugate group elements are isomorphically mapped to conjugate permutations reflecting the action of those elements on the group itself through composition. Thus conjugate elements fulfill similar functions within the group. As an example, you might review our numerous analyses of the group D_3, comparing the functions of conjugate and nonconjugate elements in the various intragroup relations.

Problem

(a) Display a realization of the order-6 group D_3 as a subgroup of the order-6! $= 144$ group S_6 by Cayley's theorem. (Note that all of S_3 realizes D_3 due to their isomorphism.) (b) Express the elements of this subgroup in cycle form and compare conjugate elements. (c) Express these elements in the form of transpositions and find their parity. Verify that all the even permutations realize the nontrivial invariant subgroup of D_3.

Symmetry

In this chapter we study the general theory of symmetry. We learn what symmetry is and see how it is expressed in several kinds of system. Mathematical concepts presented in the preceding chapters are used extensively. Our road to symmetry starts with the concept of a state space of a system and proceeds through the concepts of transformation, transformation group, equivalence relation for a state space, symmetry transformation, and symmetry group. The quantification of symmetry is discussed, and a special section is devoted to symmetry of quantum systems.

Except for the necessary mathematical concepts, presented in the preceding chapters, the general theory of symmetry is developed from scratch. For most of the examples and problems the mathematical background required for the examples and problems of the preceding chapters more than suffices, while some basic familiarity with physical phenomena is also desirable. However, for Section 3.7, Quantum Systems, and for some of the examples and problems in the other sections a good understanding of the Hilbert-space (David Hilbert, German mathematician, 1862–1943) formulation of quantum theory and of some rather advanced physics is a prerequisite. This material is definitely not for the "general reader," who is advised to skip it.

3.1. SYSTEM, STATE

We start by presenting the concepts of system, subsystem, state, and state space. These concepts are purposely not sharply defined and assumed to be intuitively understood, in order to allow the widest possible applicability of the following development.

A *system* is whatever we investigate the properties of. We are purposely being vague and general here, so as to impose no limitations on the possible objects of our interest. A system might be abstract or concrete, microscopic or macroscopic, static or dynamic, finite or infinite. Anything (and not only things) can be a system!

A *subsystem* is a system wholly contained, in some sense, within a system. Again, vagueness and generality reign.

A *state* of a system is a possible condition of the system. Here too we are saying very little. We can add, though, that a system might have a finite or (denumerably or nondenumerably) infinite number of states, and that the same system might have different kinds or numbers of states, depending on how it is being considered.

A *state space* of a system is the set of all states of the same kind, where "the same kind" may be interpreted in any useful way.

Consider some examples.

1. Let the system be a given amount of a certain pure gas. Considered microscopically, the states of the system are characterized by the positions and momenta of all the molecules (assuming spinless, structureless, point particles). The set of all allowable positions and momenta for all molecules forms a state space of the system. Macroscopic states, assuming that the gas is confined and in equilibrium, are described by any two quantities among, say, pressure, volume, and temperature. So a state space might be the set of all pressures and temperatures. In both cases, microscopic and macroscopic, the number of states is infinite, though "more so" in the microscopic case.

2. Take as the system a plane figure. States of such a system are described by shape, size, location, orientation, and even color, texture, and so on, and are infinite in number.

3. Or to be more explicit, let the system be a plane figure of given shape and size, lying in a given plane, and with one of its points fixed in the plane. Ignore color, texture, and the like. States of this system are characterized by a single angle specifying its rotational orientation in the plane and by its *handedness* (also called *chirality*), i.e., which of its two mirror-twin versions (such as left and right hands) is appearing.

4. Consider the system of a ball lying in any one of three depressions in the sand. This system has three states.

5. Consider the system of three balls lying in three depressions. States are described by specifying which ball is in which depression. The state space of this system consists of six states, the six possible arrangements of three balls.

6. The system might be any quantum system. Its states are quantum

states, characterized by a set of quantum numbers, the eigenvalues of a complete set of commuting Hermitian operators (after Charles Hermite, French mathematician, 1822–1901). State space is postulated to be a Hilbert space, in which states are represented by vectors (actually rays). Quantum systems are treated in more detail in Section 3.7.

Problems

1. Considering only properties such as number, color, and suit, describe various state spaces for a pair of playing cards drawn from the same deck.

2. You have a set of five different flags and three flagpoles, each of which may either be empty or fly one of the flags. How many states does this system have, if you are interested only in which flags are flying and do not care on which flagpoles? How many states if the positions of the flags are important?

3. The system is a fully unfurled flag flying from a flagpole. What are the system's states?

4. What are the states of a regular pentagon of given size, lying in a given plane, with its center fixed at a given point? (It might help to label *two* points.)

5. What are the states of a straight line lying in a given plane, say the xy-plane? Express your result analytically.

6. You have a mixture of the gases H_2, Cl_2, and HCl in a closed container of constant volume with given numbers of hydrogen and chlorine atoms. Describe the system's microscopic states. Assume that the mixture is in equilibrium, meaning dynamical equilibrium, since the reactions $H_2 + Cl_2 \rightleftharpoons 2HCl$ are continually taking place. The relative rates of the combination and dissociation reactions are directly affected by temperature, pressure, and densities. Describe the macroscopic states of the system.

7. Describe the most inclusive state space for a system consisting of a free proton and a free neutron, where the particles have spin (i.e., a half unit of intrinsic angular momentum apiece) but are otherwise considered as being classical and nonrelativistic point particles.

8. Describe the most inclusive state space for a system of two free nucleons (a nucleon is a proton or a neutron), where the total energy of the system equals some given value. The particles are considered as being point particles with spin, and the quantum rule for relative

spin orientation is assumed to hold; otherwise the system is classical. Use relativistic kinematics.

9.   What is the space of internal states for a hydrogen atom?

3.2.   TRANSFORMATIONS, TRANSFORMATION GROUP

A *transformation* of a system is a mapping of a state space of the system into itself; i.e., every state has an image, which is also a state; more than one state might have the same image, but not every state must necessarily serve as an image in the mapping. We denote a transformation T by

$$u \xrightarrow{T} v \quad \text{or} \quad v = T(u)$$

(arrow notation or function notation), where u and v represent states of the system. State v is the image of state u under transformation T. State u runs over all states of a state space. If the number of states of a state space is finite, a transformation might be expressed as a two-column table. Otherwise it must be expressed as a general rule.

Transformations are composed by consecutive application, i.e., the composition of two transformations is defined as the result of applying one transformation to the result of the other. For transformations T_1 and T_2 one possible composition is $T_2 T_1$, denoting the result of first applying T_1 and then T_2:

$$
\begin{aligned}
w &= T_2(v) \\
&= T_2(T_1(u)) \\
&\overset{\text{def}}{=} (T_2 T_1)(u).
\end{aligned}
$$

The function notation practically "forces" us to specify consecutively applied transformations from right to left, and we adhere to this convention. The other possible composition of T_1 and T_2 is, of course, $T_1 T_2$, where T_2 is applied first, followed by T_1. For composition of transformations to be meaningful, the transformations must act in the same state space. This will always be tacitly assumed. As defined, the composition of any two transformations is a mapping of a state space into itself and is therefore also a transformation. Thus the set of transformations of a system for any of its state spaces is closed under the composition of consecutive application.

In general

$$T_2 T_1 \neq T_1 T_2 .$$

However, a pair of transformations T_1, T_2 such that

$$T_2 T_1 = T_1 T_2,$$

i.e.,

$$(T_2 T_1)(u) = (T_1 T_2)(u)$$

for all states u of a state space, are said to *commute*.

Composition of transformations by consecutive application is easily shown to be associative, i.e.,

$$T_3(T_2 T_1) = (T_3 T_2)T_1$$

for all transformations T_1, T_2, T_3, so that the notation $T_3 T_2 T_1$ is unambiguous. Using the definition of composition, the transformation on the left-hand side of this equation acting on arbitrary state u is

$$(T_3(T_2 T_1))(u) = T_3((T_2 T_1)(u))$$
$$= T_3(T_2(T_1(u))).$$

Similarly, the right-hand side gives

$$((T_3 T_2)T_1)(u) = (T_3 T_2)(T_1(u))$$
$$= T_3(T_2(T_1(u))),$$

which is the same. This proves associativity. In arrow notation, which is more awkward here, the same proof is

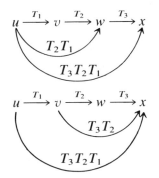

The mapping I of every state of a state space to itself,

$$u \xrightarrow{\ I\ } u \quad \text{or} \quad u = I(u)$$

for all states of a state space, is called the *identity transformation*. Indeed it acts as the identity under composition by consecutive application,

$$TI = IT = T$$

for all T, since for arbitrary state u

$$(TI)(u) = T(I(u)) = T(u),$$
$$(IT)(u) = I(T(u)) = T(u).$$

In arrow notation this is

Among the transformations of a system there are those which are one-to-one and onto (bijective), i.e., every state of a state space serves as an image of the mapping and is the image of only a single state. For those transformations, and only for those, can inverses be defined. For any transformation T that is one-to-one and onto,

$$u \xrightarrow{T} v \quad \text{or} \quad v = T(u),$$

the inverse, denoted T^{-1}, is defined as the mapping obtained by reversing the sense of the T mapping,

$$v \xrightarrow{T^{-1}} u \quad \text{or} \quad u = T^{-1}(v).$$

This was explained in Section 1.2.

Under composition by consecutive application of transformations, T and T^{-1} are mutual inverses, since their compositions in both orders are the identity transformation,

$$T^{-1}T = TT^{-1} = I.$$

This is because

$$u \xrightarrow{T} v \xrightarrow{T^{-1}} u \quad \text{or} \quad (T^{-1}T)(u) = T^{-1}(T(u))$$
$$\underbrace{\qquad}_{T^{-1}T} \qquad\qquad\qquad\qquad = T^{-1}(v)$$
$$= I \qquad\qquad\qquad\qquad\qquad = u$$
$$\qquad\qquad\qquad\qquad\qquad = I(u)$$

for all states u, and

$$v \xrightarrow{T^{-1}} u \xrightarrow{T} v \quad \text{or} \quad \begin{aligned}(TT^{-1})(v) &= T(T^{-1}(v)) \\ &= T(u) \\ &= v \\ &= I(v)\end{aligned}$$

$$\underset{= I}{\underbrace{TT^{-1}}}$$

for all states v.

It should be clear by now that we are discovering a group. The set of all invertible transformations of any state space of a system forms a group under composition of consecutive application, called a *transformation group* of the system. (1) The set is closed under composition, since the composition of invertible transformations is an invertible transformation. Indeed, the composition of one-to-one (injective) transformations is a one-to-one transformation, and, if every state of a state space serves as an image for each of two transformations, every state will serve as an image for their composition [or, in short, the composition of onto (surjective) transformations is an onto transformation]. (2) We proved associativity. (3) The identity transformation is invertible. (4) Existence of inverses holds by definition. Note that a transformation group of a system is not unique, i.e., *the* transformation group of a system is not defined in general, as the same system might have different transformation groups for different state spaces.

It should be pointed out that a transformation that is one-to-one and onto (bijective), i.e., invertible, is a permutation of state space. If such a transformation is represented by a two-column table, the list of states in the image column will be just a permutation of the list of states in the object column. Thus the transformation group of any finite state space containing n states is isomorphic with the symmetric group (not to be confused with symmetry group, which we study in Section 3.5) of degree n, S_n. For infinite state spaces the idea is the same, although perhaps not so clear intuitively.

Examples include the following.

1. Consider the macroscopic equilibrium states for the system of a given amount of a certain confined pure gas. Since p, V, θ (pressure, volume, absolute temperature) are related by an equation of state, we are free to vary any two of them, as long as p, V, $\theta > 0$ (assuming an ideal gas for simplicity). Thus all transformations of the system are of the form

$$\begin{cases} p \rightarrow p' = f(p,\theta), \\ \theta \rightarrow \theta' = g(p,\theta) \end{cases}$$

for arbitrary positive functions $f(p,\theta)$, $g(p,\theta)$. For arbitrary invertible positive functions whose range of values is all the positive numbers we obtain the (infinite-order) transformation group of the system for its macroscopic state space.

2.   Considering microscopic states for the system of a given amount of a certain pure gas, each molecule (all of which are identical, since the gas is pure) has three degrees of freedom (assuming spinless, structureless, point particles), so its state is characterized by three coordinates, x, y, z, and three components of momentum, p_x, p_y, p_z. All coordinates and all components of momentum for all molecules can be varied freely. Thus all transformations of the system are of the form

$$\begin{cases} x_i \rightarrow x_i' = X_i(x_j, y_j, z_j, p_{xj}, p_{yj}, p_{zj}), \\ y_i \rightarrow y_i' = Y_i(x_j, y_j, z_j, p_{xj}, p_{yj}, p_{zj}), \\ z_i \rightarrow z_i' = Z_i(x_j, y_j, z_j, p_{xj}, p_{yj}, p_{zj}), \\ p_{xi} \rightarrow p_{xi}' = P_{xi}(x_j, y_j, z_j, p_{xj}, p_{yj}, p_{zj}), \\ p_{yi} \rightarrow p_{yi}' = P_{yi}(x_j, y_j, z_j, p_{xj}, p_{yj}, p_{zj}), \\ p_{zi} \rightarrow p_{zi}' = P_{zi}(x_j, y_j, z_j, p_{xj}, p_{yj}, p_{zj}), \end{cases}$$

for $i, j = 1, \ldots, N$, where N is the number of molecules, with arbitrary functions $X_i(\ )$, $Y_i(\ )$, $Z_i(\ )$, $P_{xi}(\ )$, $P_{yi}(\ )$, $P_{zi}(\ )$. For arbitrary invertible functions whose range of values is all the real numbers we obtain the (infinite-order) transformation group of the system for its microscopic state space.

3.   For a plane figure of given shape and size, lying in a given plane, with one of its points fixed in the plane, the angle of rotational orientation in the plane, γ, and its handedness can be varied freely. Thus all transformations of this system are

$$\gamma \rightarrow \gamma' = g(\gamma),$$

where the values of otherwise arbitrary $g(\gamma)$ are nonnegative and less than 360° [These are orientation-dependent rotations within the given plane about the fixed point by angle $g(\gamma) - \gamma$.] and reflections through all two-sided mirrors perpendicular to the given plane and passing through the fixed point. Restricting the functions $g(\gamma)$ to being invertible and having all nonnegative angle values less than 360°, we obtain, including the reflections, a transformation group (of infinite order) of the system. If, however, the transformation

$$\gamma \rightarrow \gamma' = g(\gamma)$$

is understood, not as giving the new orientation of the whole figure for every possible original orientation, but rather as giving the new orientation of each point of the figure for *every* possible original orientation of the point, whatever the state of the figure as a whole, as, in fact, is the usual interpretation (see the following section), then for the figure to retain its shape all its points must rotate through the same angle, and the functions $g(\gamma)$ are immediately restricted to the form

$$g(\gamma) = \gamma + \alpha.$$

4.   For the system of one ball and three depressions the transformation group is the group of permutations of the three states of its state space, the symmetric group of degree 3, S_3 (of order 6).

5.   For the system of three balls in three depressions the transformation group is the group of permutations of the six states of its state space, the symmetric group of degree 6, S_6 (of order 720).

6.   For any quantum system the set of all transformations is all the operators on its Hilbert space of states. The subset of all invertible operators constitutes the transformation group. However, general operators are not considered for quantum systems, but only linear and antilinear operators. The transformation group is correspondingly the subset of all invertible linear and antilinear operators. See Section 3.7 for a more detailed treatment of quantum systems. [Recall that an antilinear operator A is characterized by the properties

$$A(|u\rangle + |v\rangle) = A|u\rangle + A|v\rangle,$$

$$Az|u\rangle = \bar{z}A|u\rangle$$

for all vectors $|u\rangle$ and $|v\rangle$ and all complex numbers z, where $\bar{z}$ is the complex conjugate of z.]

Problem

Describe the transformation group for each state space of the problems of the preceding section.

3.3.   TRANSFORMATIONS IN SPACE, TIME, AND SPACE-TIME

Although we defined transformations in a very general way, and, as we saw in the examples, transformations can be very diverse indeed, transformations in space, time, and space-time are especially important, since the states of many systems are described in spatiotemporal terms. We therefore devote a section to a brief summary of some of the more important transformations of these kinds. We use Cartesian coordinates (x,y,z) (after René Descartes, French philosopher, mathematician, and scientist, 1596–1650) for space and Minkowskian coordinates (x,y,z,t) (Hermann Minkowski, Russian mathematician, 1864–1909) for spacetime. First consider a number of spatial transformations.

The transformation of *spatial displacement* (or *spatial translation*) maps all points of space to image points that are the same distance away and in the same direction. Spatial displacements are invertible transformations.

$$\begin{cases} x \rightarrow x' = x + a \\ y \rightarrow y' = y + b \\ z \rightarrow z' = z + c \end{cases} \qquad \begin{cases} x' \rightarrow x = x' - a \\ y' \rightarrow y = y' - b \\ z' \rightarrow z = z' - c \end{cases}$$

The *rotation* transformation maps all points of space to image points found by rotating about a fixed common axis by a common angle. Rotations are invertible. If the axis of rotation is taken as the z-axis and the rotation angle is α (in the positive sense, from the positive x-axis toward the positive y-axis), we have

$$\begin{cases} x \rightarrow x' = x \cos\alpha - y \sin\alpha \\ y \rightarrow y' = x \sin\alpha + y \cos\alpha \\ z \rightarrow z' = z \end{cases} \qquad \begin{cases} x' \rightarrow x = x' \cos\alpha + y' \sin\alpha \\ y' \rightarrow y = -x' \sin\alpha + y' \cos\alpha \\ z' \rightarrow z = z' \end{cases}$$

See [M34] and [M33].

The *plane reflection* (or *plane inversion*) transformation is the transformation of reflection through a fixed two-sided plane mirror, the reflection plane. The image of any point is found by dropping a perpendicular from the point to the reflection plane and continuing the line on for the same distance on the opposite side of the plane. The image point is located at the end of this line segment. Plane reflections are invertible. If the reflection plane is taken as the xy-plane, we have

$$\begin{cases} x \rightarrow x' = x \\ y \rightarrow y' = y \\ z \rightarrow z' = -z \end{cases} \qquad \begin{cases} x' \rightarrow x = x' \\ y' \rightarrow y = y' \\ z' \rightarrow z = -z' \end{cases}$$

The transformation of *line inversion* (or *line reflection*) is inversion through a fixed straight line, the inversion line. The image of any point is the point at the end of the line segment constructed by dropping a perpendicular from the original point to the inversion line and continuing the perpendicular on for the same distance. Line inversions are invertible. If the inversion line is taken as the z-axis, we have

$$\begin{cases} x \rightarrow x' = -x \\ y \rightarrow y' = -y \\ z \rightarrow z' = z \end{cases} \qquad \begin{cases} x' \rightarrow x = -x' \\ y' \rightarrow y = -y' \\ z' \rightarrow z = z \end{cases}$$

By putting $\alpha = 180°$ in the rotation formulas above, we find that line inversion and rotation by $180°$ about the inversion line are the same transformation.

For the *point inversion* (or *point reflection* or *space inversion*) transformation the image of any point is at the end of the line segment running from the object point through a fixed point, the inversion center, and on for the same distance. Point inversions are invertible. If the inversion center is taken as the coordinate origin, we have

$$\begin{cases} x \rightarrow x' = -x \\ y \rightarrow y' = -y \\ z \rightarrow z' = -z \end{cases} \qquad \begin{cases} x' \rightarrow x = -x' \\ y' \rightarrow y = -y' \\ z' \rightarrow z = -z' \end{cases}$$

The *glide* transformation is the transformation consisting of the consecutive application of displacement parallel to a fixed plane and reflection through this plane, called a glide plane. (The two transformations can just as well be applied in reverse order, since they commute.) Glide transformations are invertible. If the glide plane is taken as the xy-plane, we have

$$\begin{cases} x \rightarrow x' = x + a \\ y \rightarrow y' = y + b \\ z \rightarrow z' = -z \end{cases} \qquad \begin{cases} x' \rightarrow x = x' - a \\ y' \rightarrow y = y' - b \\ z' \rightarrow z = -z' \end{cases}$$

The *screw* transformation is the transformation resulting from the consecutive application (in either order, since they commute) of a rotation and a displacement parallel to the rotation axis, called a screw axis. Screw transformations are invertible. If the screw axis is taken as the z-axis, we have

$$\begin{cases} x \rightarrow x' = x \cos\alpha - y \sin\alpha \\ y \rightarrow y' = x \sin\alpha + y \cos\alpha \\ z \rightarrow z' = z + c \end{cases} \qquad \begin{cases} x' \rightarrow x = x' \cos\alpha + y' \sin\alpha \\ y' \rightarrow y = -x' \sin\alpha + y' \cos\alpha \\ z' \rightarrow z = z' - c \end{cases}$$

Under a *spatial dilatation* (or *scale*) transformation the image of any point is found by moving the point away from a fixed point, the dilatation center, along the straight line connecting them, to a distance from the fixed point related to the original distance by a fixed positive factor, the dilatation factor. This transformation increases the distances between all pairs of points by the same factor. (If the dilatation factor is less than 1, all distances are actually decreased.) Dilatations are invertible. If the dilatation center is taken as the coordinate origin, we have for positive ρ

$$\begin{cases} x \rightarrow x' = \rho x \\ y \rightarrow y' = \rho y \\ z \rightarrow z' = \rho z \end{cases} \quad \begin{cases} x' \rightarrow x = x'/\rho \\ y' \rightarrow y = y'/\rho \\ z' \rightarrow z = z'/\rho \end{cases}$$

The image of any point under a *plane projection* transformation is a point in a fixed plane, the projection plane. It is the point of penetration of the perpendicular dropped from the original point to the projection plane. Projections are not invertible; they are neither one-to-one nor onto. If the projection plane is taken as the xy-plane, we have

$$\begin{cases} x \rightarrow x' = x \\ y \rightarrow y' = y \\ z \rightarrow z' = 0 \end{cases}$$

Under a *line projection* transformation the image of any point is a point on a fixed line, the projection line. The image is the point of intersection of the perpendicular dropped from the original point to the projection line. If the projection line is taken as the z-axis, we have

$$\begin{cases} x \rightarrow x' = 0 \\ y \rightarrow y' = 0 \\ z \rightarrow z' = z \end{cases}$$

Now look at three temporal transformations. All are invertible. The first is *temporal* (or *time*) *displacement* (or *translation*), under which the images of all instants are the same time interval away from the original instants:

$$t \rightarrow t' = t + d, \qquad t' \rightarrow t = t' - d.$$

The transformation of *temporal* (or *time*) *inversion* (or *reflection* or *reversal*) maps each instant to the instant that is the same time interval before a fixed instant, the central instant, as the original instant is after it, or vice versa. If the inversion central instant is taken as the temporal origin, we have

$$t \rightarrow t' = -t, \qquad t' \rightarrow t = -t'.$$

The *temporal* (or *time*) *dilatation* (or *scale*) transformation maps all instants to instants whose time intervals from a fixed instant, the central instant, are larger by a fixed positive factor, the dilatation factor, than those of the respective object instants. This transformation increases the time interval between all pairs of instants by the same factor. (If the dilatation factor is less than 1, all time intervals are actually decreased.) If the dilatation central instant is taken as the temporal origin, we have for positive σ

$$t \rightarrow t' = \sigma t, \qquad t' \rightarrow t = t'/\sigma.$$

Our first spatiotemporal transformation is the *Lorentz transformation* (Hendrik Antoon Lorentz, Dutch physicist, 1853–1928) (or *boost*), which maps all events to the events whose coordinates are the same as those that an observer moving with constant rectilinear velocity would assign to the original events with respect to his or her rest frame (the Minkowskian frame with respect to which he or she is at rest). Lorentz transformations are invertible. These transformations are an essential ingredient of Albert Einstein's (American physicist, 1879–1955) special theory of relativity. If the observer is moving in the negative x direction with velocity v, such that $-c < v < c$, where c is the speed of light, and if his or her Minkowskian coordinate axes are parallel to ours and his or her origin coincides with ours at time $t = 0 = t'$, we have

$$\begin{cases} x \rightarrow x' = \gamma(x + vt) \\ y \rightarrow y' = y \\ z \rightarrow z' = z \\ t \rightarrow t' = \gamma(t + vx/c^2) \end{cases} \qquad \begin{cases} x' \rightarrow x = \gamma(x' - vt') \\ y' \rightarrow y = y' \\ z' \rightarrow z = z' \\ t' \rightarrow t = \gamma(t' - vx'/c^2) \end{cases}$$

where

$$\gamma = 1/\sqrt{(1 - v^2/c^2)}.$$

Our other spatiotemporal transformation is the *Galilei transformation* (Galileo Galilei, Italian astronomer, mathematician, and physicist, 1564–1642) (or *nonrelativistic boost*), defined like the Lorentz transformation, except that the mathematical limit of $c \rightarrow \infty$ (or $v/c \rightarrow 0$) is taken.

With the same assumptions as above we have

$$\begin{cases} x \rightarrow x' = x + vt \\ y \rightarrow y' = y \\ z \rightarrow z' = z \\ t \rightarrow t' = t \end{cases} \qquad \begin{cases} x' \rightarrow x = x' - vt' \\ y' \rightarrow y = y' \\ z' \rightarrow z = z' \\ t' \rightarrow t = t' \end{cases}$$

Problems

1. Except for the projection transformations, all transformations and sets of transformations presented in this section generate or form groups. Every set of transformations containing continuous parameters forms a group, called a continuous group, where certain restrictions might apply to the ranges of the parameters. Prove this for each such set of transformations except the glide and find the restrictions, if any. Prove that each reflection or inversion transformation generates an order-2 group. (Groups of finite or denumerably infinite order are called discrete groups.) The set of glide transformations does not form a group, because it is not closed and the identity is missing. Prove, however, that the set of glide transformations with fixed glide plane generates a group.

2. Some of the transformations presented in this section can be generalized easily. Generalize the following transformations as indicated: (a) rotation about arbitrary axis parallel to the z-axis; (b) plane reflection with arbitrary reflection plane; (c) line inversion with arbitrary inversion line; (d) point inversion with arbitrary inversion center; (e) glide with arbitrary glide plane; (f) spatial dilatation with arbitrary dilatation center; (g) plane projection with arbitrary projection plane; (h) line projection with arbitrary projection line; (i) temporal inversion with arbitrary central instant; (j) temporal dilatation with arbitrary central instant; (k) Lorentz transformation with arbitrary direction of velocity v; (l) Galilei transformation with arbitrary direction of velocity v.

3. Many of the sets of transformations presented in this section can be combined and expanded to form larger groups than those discussed in Problem 1. This is especially simple when the transformations commute, such as spatial displacements parallel to the z-axis and rotations about that axis, generating the group of screw transformations. When the transformations do not commute, one must be careful. Prove that each of the following sets of transformations forms or generates a group: (a) rotations about the z-axis and spatial dila-

tations (with center at the origin); (b) spatial displacements parallel to the xy-plane and rotations about all axes parallel to the z-axis; (c) spatial displacements and spatial dilatations with all dilatation centers; (d) spatial displacements, plane reflections with reflection plane parallel to the xy-plane, and glides with glide planes parallel to the xy-plane; (e) Lorentz transformations with velocity v parallel to the x-axis, spatial displacements parallel to the x-axis, and temporal displacements; (f) like (e) but with Galilei transformations instead of Lorentz transformations.

3.4. STATE EQUIVALENCE

We now consider the possibility of defining an equivalence relation for a state space of a system. This is, we recall (Section 1.4), any relation, denoted $\equiv$, that holds between any two states and satisfies the following three conditions:

1. *Reflexivity*.

$$u \equiv u$$

for all states u.

2. *Symmetry*.

$$u \equiv v \Leftrightarrow v \equiv u$$

for all states u, v.

3. *Transitivity*.

$$u \equiv v, \; v \equiv w \Rightarrow u \equiv w$$

for all states u, v, w.

Recall that any subset of states of a state space—such a subset is called a subspace—such that all the states comprising it are equivalent to each other and no other state is equivalent to any state of the subset is called an equivalence class, or, in our case, an equivalence subspace. Recall also that an equivalence relation brings about a decomposition of a state space into equivalence subspaces. (See Section 2.1.)

Some examples of state equivalence follow.

1. If microscopic states of the gas of previous examples are being considered, while only the macroscopic properties of the gas are really

of interest, any two microscopic states corresponding to the same pressure, volume, and temperature can be taken as equivalent. Actually there is an infinite number of microscopic states corresponding to any given macroscopic state and thus forming an equivalence subspace. As for the infinity of microscopic states that do not correspond to any equilibrium macroscopic state, we can define all of these as equivalent to each other.

2.   Considering only macroscopic states of the gas, for some purposes states with the same temperature, for example, might be equivalent. This would occur if the gas served solely as a heat sink. Then all states with the same temperature would form an equivalence subspace.

3.   If we take an equilateral triangle for the plane figure of given shape and size, lying in a given plane, with one of its points fixed in the plane, as far as appearance is concerned any state of the triangle will be equivalent to the two states obtained from it by rotations by 120° and by 240° about the axis through its center and perpendicular to its plane and to the three states obtained from it by reflection through the three planes containing this axis and a median. Thus the state space of this system decomposes into equivalence subspaces of six equivalent states each.

4.   If this triangle has one vertex marked or if it is deformed into an isosceles triangle, then the states obtained from any given state by rotations by 120° and 240° will no longer be equivalent to the original state with respect to appearance. And of the reflections only reflection through the plane containing the median of the different vertex and perpendicular to the plane of the triangle will still produce an equivalent state. Now the state space of the system decomposes into equivalence subspaces of only two states each.

5.   If yet another vertex of the equilateral triangle is marked, and differently from the first, or if the isosceles triangle is further deformed into a scalene triangle, then all equivalence among states, with regard to appearance, will disappear.

6.   In the system of one ball and three depressions, if the three depressions are adjacent to one another and the scene is viewed from a distance, as far as appearance is concerned all three states of the system will be equivalent.

7.   If only two of the depressions are adjacent to one another and the third is separated from those two, as far as appearance is concerned two of the states will be equivalent to each other and the third will not be equivalent to either of those, so that state space will decompose into two equivalence subspaces, one containing two states and the other containing one state.

8.   If the system is viewed from close by, no state will look like another, and there will be no equivalence with regard to appearance.

9.   In the system of three balls in three depressions, if all three balls look the same, then as far as appearance is concerned all six states of the system will be equivalent to each other.

10.   If two balls look alike and the third is distinct, each state of the system will be equivalent to the state obtained from it by interchanging the similar balls. Thus the state space of six states will decompose into three equivalence subspaces of two states each.

11.   If all three balls look different, there will be no equivalence among states with regard to their appearance.

12.   Since quantum systems are especially important, we discuss quantum state equivalence and its consequences separately in Section 3.7.

Problem

For each of the problems of Section 3.1 impose the state-equivalence relation specified here on the state space(s) of the system, describe equivalence subspaces, and describe the decomposition of the state space(s) into equivalence subspaces: (a) In Problem 1 cards of the same numerical value are equivalent; (b) in Problem 2 discuss the equivalence relation imposed there; (c) in Problem 3 flag positions of the same height are equivalent; (d) in Problem 4 states that look the same are equivalent; (e) in Problem 5 the equivalence relation is parallelism; (f) in Problem 6 equivalent states have the same amount of HCl; (g) in Problems 7 and 8 equivalent states are states that would be the same if protons and neutrons were the same; (h) in Problem 9 equivalent states have the same energy, ignoring spin-spin and spin-orbit interactions.

3.5.   SYMMETRY TRANSFORMATIONS, SYMMETRY GROUP

For a state space equipped with equivalence relation some of the transformations might leave equivalence subspaces invariant, i.e., there might be transformations that map every state to an image state equivalent to the object state. Such a transformation is called a *symmetry transformation* for the equivalence relation. Thus the defining property of a symmetry transformation S is

$$u \xrightarrow{\ S\ } v \equiv u \quad \text{or} \quad S(u) = v \equiv u$$

for all states u.

The set of all invertible symmetry transformations of a state space of a system for an equivalence relation forms a group, a subgroup of the transformation group, called the *symmetry group* of the system for the equivalence relation. (Do not confuse symmetry group with symmetric group!) This is seen as follows.

1. Closure follows from transitivity of the equivalence relation, as either of the following diagrams proves.

$$
\begin{cases}
u \equiv v = S_1(u), \\[2mm]
v \equiv w = S_2(v) = S_2(S_1(u)) \\[2mm]
\quad\quad \overset{\text{def}}{=} (S_2 S_1)(u) \\[2mm]
u \equiv w = (S_2 S_1)(u)
\end{cases}
$$

for all states u and v and symmetry transformations S_1 and S_2. Thus for all symmetry transformations S_1 and S_2 their composition $S_2 S_1$ is also a symmetry transformation, and by similar reasoning so is $S_1 S_2$.

2. Associativity holds for composition by consecutive application, as proved in Section 3.2.

3. The identity transformation is a symmetry transformation. This follows from reflexivity of the equivalence relation:

$$
u \overset{I}{\longrightarrow} u \equiv u \quad \text{or} \quad I(u) = u \equiv u
$$

for all states u.

4. The inverse of any invertible symmetry transformation is also a symmetry transformation. If S is an invertible symmetry transformation, so that

$$
u \overset{S}{\longrightarrow} v \equiv u \quad \text{or} \quad S(u) = v \equiv u
$$

for all states u, then by the symmetry property of the equivalence relation

$$
v \overset{S^{-1}}{\longrightarrow} u \equiv v \quad \text{or} \quad S^{-1}(v) = u \equiv v
$$

for all states v. Thus S^{-1} is indeed a symmetry transformation.

This proves that the set of all invertible symmetry transformations of a state space of a system for an equivalence relation forms a group, a subgroup of the transformation group of the state space. In general a system might have different symmetry groups for its different state spaces and for the different equivalence relations that might be defined for them. If no two states are equivalent, the symmetry group will consist only of the identity transformation and the system will be *asymmetric*. If all states are equivalent to each other, the symmetry group will be the transformation group itself.

This line of reasoning, by which any equivalence relation determines a subgroup of the transformation group, can be reversed to allow any subgroup of the transformation group to determine an equivalence relation. Given such a subgroup, the appropriate equivalence relation is simply: State u is equivalent to state v if and only if some element T of the subgroup performs

$$u \xrightarrow{\;T\;} v \quad \text{or} \quad v = T(u).$$

That this relation is indeed an equivalence is seen as follows:

1. The reflexivity property follows from the identity transformation's belonging to the subgroup.
2. The symmetry property follows from the existence of inverses for the subgroup.
3. Transitivity follows from closure.

Such a subgroup might or might not be the symmetry group for the equivalence relation it determines in this way. Clearly all elements of the subgroup are symmetry transformations for the equivalence relation it determines. There might, however, be additional invertible symmetry transformations. If not, the subgroup is the symmetry group. In any case it is a subgroup of the symmetry group.

Consider some examples of symmetry groups.

1.   In the case of the gas, where microscopic states corresponding to the same macroscopic state are considered equivalent to each other, the (infinite-order) symmetry group, transforming microscopic states into equivalent states only, is a subgroup of the (infinite-order) transformation group.

2.   If macroscopic states of the gas having the same temperature are considered equivalent, we will have the constraint $\theta(p, V) = \text{const}$ (isotherms), and all symmetry transformations will be of the form

$$p \rightarrow p' = f(p, \theta)$$

for arbitrary positive functions $f(p,\theta)$, a subset of all transformations of this state space as presented in Section 3.2. For arbitrary invertible positive functions $f(p,\theta)$ whose range of values is all positive numbers we obtain the (infinite-order) symmetry group, a subgroup of the (infinite-order) transformation group.

3.   The symmetry group of the equilateral triangle discussed in Section 3.4 is the group consisting of the identity transformation, rotations about the axis through the center of the triangle and perpendicular to its plane by 120° and 240°, and reflections through each of the three planes containing the rotation axis and a median, the order-6 group D_3. Refer to Fig. 3.1. This group is a subgroup of the transformation group of the system, which is the same as the transformation group of any plane figure of given shape and size, lying in a given plane, with one of its points fixed in the plane, and was presented in Section 3.2.

4.   For the equilateral triangle with one vertex marked or for an isosceles triangle the symmetry group consists only of the identity transformation and reflection through the single plane containing the median of the distinct vertex and perpendicular to the plane of the triangle, the order-2 group C_2. See Fig. 3.2. This group is a subgroup of the transformation group of the system and is also a proper subgroup of the symmetry group of the equilateral triangle.

5.   For the equilateral triangle with two vertices marked differently or for a scalene triangle the symmetry group consists only of the identity transformation. These systems are asymmetric. This trivial symmetry group is a subgroup of the transformation group of the system and is also a proper subgroup of the symmetry group of the singly marked equilateral triangle or the isosceles triangle and thus also a proper subgroup of the symmetry group of the equilateral triangle.

6.   For a similar example, replace the equilateral triangle of the preceding examples by a square. The symmetry group of this system for the

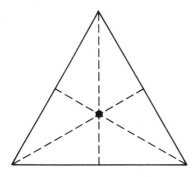

Figure 3.1.   Equilateral triangle. Center is point of intersection of axis of threefold rotation symmetry. Medians are lines of intersection of planes of reflection symmetry.

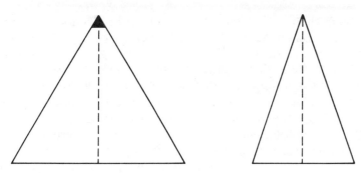

Figure 3.2. Equilateral triangle with one vertex marked and isosceles triangle. Median of distinct vertex is line of intersection of plane of reflection symmetry.

equivalence relation of identical appearance consists of the identity transformation, rotations by 90°, 180°, and 270° about the axis through the center of the square and perpendicular to its plane, reflections through each of the two planes containing the axis and a diagonal, and reflections through each of the two planes containing the axis and parallel to a pair of edges, the order-8 group D_4. Refer to Fig. 3.3. This group is a subgroup of the transformation group of the system (the same as that of any plane figure of given shape, etc.).

We take this opportunity to give the definition of the general *dihedral group D_n*. In generalization of the symmetry groups of the equilateral triangle and the square, the dihedral group D_n is the symmetry group of the regular *n*-sided plane polygon. It consists of the identity transformation, rotations about the axis through the center of the polygon and perpendicular to its plane by $360°/n$, $2 \times 360°/n$, $3 \times 360°/n$, ..., $(n - 1) \times 360°/n$, and reflections through each of the *n* planes containing the axis and a vertex or the center of a side (or both). D_n is of order $2n$.

7.   If the square is squeezed into a rectangle, its symmetry group will reduce to the identity transformation, rotation by 180° about the axis through the center of the rectangle and perpendicular to its plane (or inver-

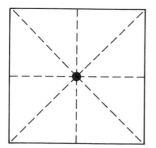

Figure 3.3. Square. Lines of intersection of planes of reflection symmetry are indicated. Center is point of intersection of fourfold rotation symmetry axis.

sion through the axis, which is the same), and reflections through each of two planes containing the axis and parallel to a pair of edges, the order-4 group D_2. Refer to Fig. 3.4. This is a subgroup of the transformation group of the system (that of any plane figure of given shape, etc.) and is also a proper subgroup of the symmetry group of the square.

8.   If in the system of one ball and three depressions all three depressions are adjacent so that all three states look the same from a distance, then all states will be equivalent, and the symmetry group will be the transformation group itself, the order-6 group S_3. The symmetry group in this case happens to be isomorphic with the symmetry group of the equilateral triangle of given size, etc., D_3.

9.   If only two of the depressions are adjacent, so that only two of the three states are equivalent to each other, the symmetry group will consist of the identity transformation and the transposition interchanging the two equivalent states, the order-2 group S_2. This group is a subgroup of the transformation group of the system, S_3, which is also the symmetry group of the system with all depressions adjacent. In addition, S_2 is isomorphic with the symmetry group of the isosceles triangle (or equilateral triangle with one vertex marked) of given size, etc., C_2.

10.   If all three depressions are separated and no states are equivalent, the symmetry group will consist only of the identity transformation. This is a subgroup of the transformation group and is also a proper subgroup of the symmetry group of the system with two adjacent depressions and thus also a proper subgroup of the symmetry group of the system with three adjacent depressions.

11.   If in the system of three balls in three depressions all three balls look the same, then, since all states are equivalent, the symmetry group will be the transformation group itself, the order-720 group S_6.

12.   If only two balls look alike and the third is distinct, the symmetry group will be the direct product of the permutation groups of the individual

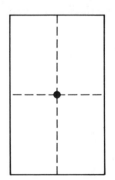

Figure 3.4.   Rectangle. Lines of intersection of planes of reflection symmetry are indicated. Center is point of intersection of twofold rotation symmetry axis.

equivalence subspaces, the direct product of the symmetric group of degree 2 with itself twice, $S_2 \times S_2 \times S_2$, an order-8 group. This group is a subgroup of the transformation group of the system, S_6, which is also the symmetry group of the system with three similar balls.

13.   If all three balls look different, the symmetry group will consist only of the identity transformation. This is a subgroup of the transformation group and is also a proper subgroup of the symmetry group of the system with two similar balls and thus also a proper subgroup of the symmetry group of the system with three similar balls.

Note that in each of examples 8 to 13 the state space is finite and we found the symmetry group exactly according to its definition. In each of examples 1 to 12, however, the state space is infinite and decomposes into an infinite number of equivalence subspaces. What we found is the transformation group of each equivalence subspace rather than the symmetry group of the entire state space. The latter is the infinite direct product of the transformation groups of the individual equivalence subspaces, which is so awkward that the normal procedure is to do just what we did.

14.   Symmetry groups for quantum systems are discussed in Section 3.7.

Problems

1.   Prove in detail that a subgroup of the transformation group of a state space defines an equivalence relation for the state space, as described in this section, and that the defining subgroup is a subgroup of the symmetry group for the equivalence relation it defines.

2.   Prove that the symmetry groups of the equilateral triangle and square, of given size and given location in a given plane, with identical-looking states considered equivalent, are indeed D_3 and D_4, respectively, as claimed in the text.

3.   Prove that the symmetry group of the system of three balls in three depressions with two of the balls similar is $S_2 \times S_2 \times S_2$, as claimed in example 12.

4.   For each of the Platonic solids find all the rotations, plane reflections, and point inversions that are symmetry transformations. (It is a very good idea to construct models and work with them.) Compare your results for the cube with those for the regular octahedron and your results for the regular dodecahedron with those for the regular icosahedron. In what way is the regular tetrahedron "odd man out"?

5. Find all the geometric symmetry (i.e., appearance-preserving) transformations of an infinite bathroom floor tiled with identical regular hexagons.

6. Find all the symmetry (i.e., appearance-preserving) transformations for (a) a standard chess board and (b) an infinite chess board. (*Hint:* In addition to geometric transformations consider the transformation of color interchange and its composition with other transformations.)

7. Find the symmetry group for each state space of the problems of Section 3.1 with the state-equivalence relation specified in the problem of Section 3.4.

8. As an example of the interesting byways into which symmetry can lead us, a person found that both $S_2 \times S_2 \times S_2 \times S_2 \times S_2 \times S_2 \times S_2 \times S_2$ and $S_9 \times S_9 \times S_2$ were symmetry groups of the 20 horses in his or her stable. What equivalence relations might he or she have had in mind?

3.6. QUANTIFICATION OF SYMMETRY

We now have at our disposal the means to attempt to quantify symmetry or at least set up a hierarchy, or ordering, of symmetries. A quantification of symmetry would be a way of assigning a number to each symmetry group, expressing the degree of symmetry of a system having that symmetry group. A symmetry ordering would be a way of comparing any two symmetry groups to determine which of two systems, each possessing respectively one of the two symmetry groups, has the higher degree of symmetry. A symmetry quantification is clearly a symmetry ordering, although an ordering might not go so far as to be a quantification.

There are three properties that, on the basis of our experience, we may reasonably expect of any scheme of symmetry quantification or ordering. First of all, if the symmetry group of a system is isomorphic with the symmetry group of another system (e.g., the equilateral triangle of given size, etc., and the ball and three adjacent depressions, or the isosceles triangle and the ball and two adjacent and one separated depressions), it is reasonable to consider them as having the same degree of symmetry (even though their symmetry transformations might be of entirely different character, as in the examples). Thus it is the abstract group of which the symmetry group is a realization which is of interest, rather than the symmetry group itself.

Next, if the symmetry group of a system is isomorphic with a proper subgroup of the symmetry group of another system (e.g., the isosceles triangle and the equilateral triangle, or the ball and two adjacent and one separated depressions and the ball and three adjacent depressions, or the isosceles triangle and the ball and three adjacent depressions, or the rectangle and the square), the latter system may reasonably be considered more symmetric than the former.

And finally, if a system is asymmetric—if its symmetry group consists solely of the identity transformation—it is reasonable to assign it the lowest degree of symmetry.

But how should we compare the degrees of symmetry of systems whose symmetry groups are neither isomorphic with each other nor one isomorphic with a proper subgroup of the other (e.g., the square and the equilateral triangle)? The main purpose of any symmetry quantification or ordering scheme is to answer just that question.

One possibility for a quantification scheme is to take for the degree of symmetry of a system the order of its symmetry group (or any monotonically ascending function thereof, such as its logarithm). For finite-order symmetry groups (and we do not go into infinite-order groups here) this is a perfectly satisfactory scheme in that it possesses the three properties that we demanded previously: Isomorphic groups have the same order, a proper subgroup is of lower order than its including group, and the order-1 group has the lowest order of all. It might, however, well be objected on philosophical grounds that this quantification scheme assigns equal weights to all the symmetry transformations of a symmetry group, while in fact they are not all independent, as they can be generated by repeated and consecutive applications of a minimal set of generators. We are thus led to consider the scheme whereby the degree of symmetry of a system is taken to be the minimal number of generators of its symmetry group. As for the three properties, it possesses the first; isomorphic groups have the same minimal numbers of generators. It is, however, defective with respect to the second property; a proper subgroup does not necessarily have a lesser minimal number of generators than its including group. For example, the order-2 cyclic group C_2 is a proper subgroup of the order-4 cyclic group C_4, but, being cyclic, each group can be generated by positive powers of a single element. The third property can be forced by assigning the value 0 to the order-1 group.

For comparison we list in Fig. 3.5 some of the symmetries we found in our examples along with the degrees of symmetry assigned to the systems by the two symmetry-quantification schemes considered. Note, for example, that the minimal-number-of-generators scheme assigns equal degrees of symmetry to the square and to the rectangle. The group-order scheme is clearly preferable. Other, intermediate schemes could be de-

System	Symmetry Group	Order	Minimal Number of Generators
Square	D_4	8	2
Rectangle	D_2	4	2
Equilateral triangle	$D_3 \sim S_3$	6	2
Isosceles triangle	$C_2 \sim S_2$	2	1
Ball and three adjacent depressions	$S_3 \sim D_3$	6	2
Ball and two adjacent and one separated depressions	$S_2 \sim C_2$	2	1
Asymmetric system	C_1	1	0 (defined)

Figure 3.5. Table of orders and minimal numbers of generators for various symmetry groups.

vised, giving the degree of symmetry of a system as some function of both the order and the minimal number of generators of its symmetry group.

3.7. QUANTUM SYSTEMS

This section is a concise presentation intended to place the subject of symmetry in quantum systems within the framework developed in the preceding sections of this chapter. Thus it should introduce, but by no means replace, the various discussions of the same subject found elsewhere. It is intended for readers with a good understanding of the Hilbert-space formulation of quantum theory; readers lacking this prerequisite are advised to skip it.

Quantum systems, which are but special cases of the general systems we have been discussing, are so important they deserve separate treatment. Their special nature allows us to be more specific in our discussion of their symmetry.

The state spaces of quantum systems are Hilbert spaces, and quantum states are vectors in these spaces. (Actually, quantum states are *represented* by vectors in Hilbert spaces. More accurately, they are represented by *rays,* since all complex multiples of any vector are supposed to represent the same state. But for the convenience of the present discussion we identify state with vector. For the sake of simplicity we also ignore possible complications due to coherent subspaces, which in any case can be overcome with slight additional effort.) The transformations of interest for quantum systems are linear and antilinear transformations in their Hilbert state spaces, said to be implemented by linear and antilinear operators. In the following discussion we use "operator" to mean "linear or antilinear operator." Thus the transformation group of a quantum system is the group of all invertible operators in its Hilbert space of states.

Now what properties of states of quantum systems might be utilized for the definition of equivalence relations? Consider the following:

1. *Transition Amplitudes.* Each state $|u\rangle$ is characterized by the transition amplitudes between itself and all states of its Hilbert space, $\langle u \mid v \rangle$ or $\langle v \mid u \rangle$ for all states $|v\rangle$.

2. *Norm.* Especially, each state $|u\rangle$ is characterized by its transition amplitude with itself, $\langle u \mid u \rangle$, the square of its norm.

3. *Eigenheit.* Each state $|u\rangle$ either is or is not an eigenstate of each member of any set of operators $\{O_i\}$ in its Hilbert space, and, if it is, it is characterized by its eigenvalue.

4. *Expectation Values.* Each state $|u\rangle$ gives a set of expectation values $\{\langle u \mid O_i \mid u \rangle / \langle u \mid u \rangle\}$ for any set of operators $\{O_i\}$ in its Hilbert space.

It seems reasonable to define state equivalence for a given quantum system and a given set of operators in its Hilbert state space as indistinguishability with respect to properties 1 through 4.

The search for all symmetry operators for the equivalence relation of indistinguishability with respect to (1) transition amplitudes and (2) norm is essentially reducible to the search for all operators preserving scalar products up to complex conjugation. And the standard result is the group of all unitary and antiunitary operators in the Hilbert state space of the system. See [A1]. For strict preservation of scalar products the symmetry group consists only of all unitary operators. [Recall that antiunitary (or unitary antilinear) operators are invertible antilinear operators A obeying

$$\langle u \mid A^\dagger A \mid v \rangle = \langle v \mid u \rangle = \overline{\langle u \mid v \rangle}$$

for all states $|u\rangle$, $|v\rangle$, where the bar denotes complex conjugation and the dagger denotes Hermitian conjugation.]

Now consider property 3. Equivalent states, in addition to their being related by unitary or antiunitary operators, are also required to be either all eigenstates or all not eigenstates of each operator O_i of a given set $\{O_i\}$, and, if they are all eigenstates, they must belong to the same eigenvalue. The first part of this requirement leaves as symmetry operators only those unitary and antiunitary operators commuting with the set $\{O_i\}$.

The second part in general rejects the antiunitary operators among these, since, while eigenstates related by unitary operators indeed belong to the same eigenvalue of an operator commuting with the unitary operators, a pair of eigenstates related by an antiunitary operator commuting with the operator of which they are eigenstates belong to eigenvalues that are a complex-conjugate pair. However, if the set of operators $\{O_i\}$ consists solely of Hermitian operators, which indeed is the usual case, with Hermitian operators representing measurable physical quantities, the antiunitary operators are *not* rejected. In that case a pair of eigenstates related by an antiunitary operator commuting with the Hermitian operator of which they are eigenstates belong to the same eigenvalue, since the eigenvalues of Hermitian operators are real. A very important Hermitian operator that is almost always included in the set $\{O_i\}$ is the Hamiltonian operator (after William Rowan Hamilton, British mathematician, 1805–1865), representing the energy of the system and generating the system's evolution.

That also takes care of property 4. States related by unitary operators commuting with a given set of operators $\{O_i\}$ give equal sets of expectation values. A pair of states related by an antiunitary operator commuting with the set give complex-conjugate sets of expectation values. But if $\{O_i\}$ is a set of Hermitian operators, they will give equal sets of (real) expectation values.

In summary, for a given quantum system and a given set of operators in its Hilbert space of states the symmetry group for equivalence with respect to (1) transition amplitudes, (2) norm, (3) eigenheit, and (4) expectation values is the group of all unitary operators in the Hilbert space commuting with the given set of operators. If the given set consists solely of Hermitian operators, the symmetry group is the group of all unitary and antiunitary operators commuting with the set. (But if it is desired to preserve scalar products strictly, not just up to complex conjugation, then the antiunitary operators are excluded whether the operators of the given set are Hermitian or not.)

Problems

1. Find the symmetry transformations of the harmonic oscillator, where the set $\{O_i\}$ consists of the Hamiltonian operator only.

2. Find the symmetry transformations of the nonrelativistic, spinless hydrogen atom, where the set $\{O_i\}$ consists only of the Hamiltonian operator.

3. Find the symmetry transformations of the nonrelativistic, spinless hydrogen atom in a uniform magnetic field, where the set $\{O_i\}$ consists only of the Hamiltonian operator.

4. Find the symmetry transformations of a relativistic, free particle, where the set $\{O_i\}$ consists of the mass and spin operators.

Application of Symmetry

In this chapter we study the theory and practice of the application of symmetry in science. The practice is based on the symmetry principle, stating roughly that the effect is at least as symmetric as the cause, and the theory is the proof of this principle. The proof starts with a discussion of the concept of causal relation, followed by a clarification of certain points concerning the nature of science, from which we are led to the equivalence principle, stating roughly that equivalent causes imply equivalent effects. The symmetry principle follows almost immediately from the equivalence principle. In the application of symmetry in science the symmetry principle can be used in two ways: minimalistically, to set a lower bound on the symmetry of the effect, and maximalistically, to set an upper bound on the symmetry of the cause. Numerous examples are presented. The application of symmetry in quantum systems is briefly discussed.

Except for the concepts taken from the previous chapters, the material is developed from scratch. The theory leading to the equivalence principle is unavoidably philosophical, so be forewarned. If philosophical considerations are anathema to you, you had better just memorize the equivalence principle and skip the first two sections. But why not give it a try? It might not be as bad as you fear. However, Section 4.6, Quantum Systems, requires an understanding of quantum theory and its Hilbert-space formulation. If this is lacking, the section should be skipped. It is very brief anyhow. For the examples and problems the mathematical prerequisites of the previous chapters certainly suffice, except for a few cases that require an understanding of simple ordinary differential equations. The scientific prerequisite is an elementary understanding of certain topics in classical and modern physics: mechanics, electromagnetism, quantum mechanics, special relativity, and the Rutherford experiment (Ernest Rutherford, British physicist, 1871–1937).

4.1. CAUSAL RELATION

The first step in our development of the theory of application of symmetry is to achieve a sufficient understanding of the concept of *causal relation*, also called *cause–effect relation*. It might seem that in everyday affairs there should be no difficulty in understanding this relation; the cause brings about the effect, and the effect is a result of the cause. But is the situation really so clear? For example, at the end of a concert you applaud, whereby your clapping makes a noise that expresses your appreciation. The clapping makes the noise, so the clapping is the cause and the noise is the effect, which is satisfactory. But why do you clap? You clap to make the noise; if clapping did not make that kind of noise, you would not clap. Thus your clapping is performed because of the noise. And does that not mean that the noise is the cause and the clapping the effect? Or is it the desire to produce the noise that is the cause?

Possibly in everyday affairs we can make do with such foggy concepts, but in science our concepts must be much clearer (that would do no harm in everyday affairs either!). So we will attempt to clarify the concept of causal relation to a degree of hairsplitting sufficient for our purposes.

Let us recall that a subsystem of a system is itself a system, with states, state spaces, transformation groups, and perhaps even equivalence relations and symmetry groups. The state of a system determines the state of each of its subsystems. However, the state of a subsystem does not in general determine the state of the whole system or of any other subsystem (although that might happen in certain cases).

Consider two subsystems, which we denote A and B, of an arbitrary system. Consider all states of the whole system and imagine that we make a triple list of the pairs of states of subsystems A and B that are determined by each state of the whole system for all of its states. Of course, it is possible that the same state of a subsystem will appear more than once in the resulting list. See Fig. 4.1.

Let us now look for a correlation between states of subsystem A and states of subsystem B in this list by asking the following questions: Does the same state of subsystem A always appear with the same state of subsystem B (and possibly different states of A appear with the same state of B)? Does the same state of subsystem B always appear with the same state of subsystem A (and possibly different states of B appear with the same state of A)? If the answer to either or both questions is affirm-

States of Whole System	Determine States of	
	Subsystem A	Subsystem B
a	h	p
b	i	p
c	h	q
d	j	r
e	j	s
⋮	⋮	⋮

Figure 4.1. States of two subsystems determined by state of whole system.

ative, we say that a *causal relation* (or *cause–effect relation*) exists between the two subsystems. If the first question is answered affirmatively, we say that subsystem A is a *cause subsystem* and subsystem B is an *effect subsystem*. If the second question is answered affirmatively, we say that B is a cause subsystem and A is an effect subsystem. If the answers to both questions are affirmative, each subsystem is both cause and effect. See Figs. 4.2 and 4.3.

In this manner the definition of a causal relation between subsystems is based on a correlation between their state spaces. There is not even a hint of "bringing about," "resulting," "producing," or "causing" (in the usual sense). The states of one subsystem and those of the other are connected through the fact that they are determined by the states of the same whole system, and this connection is the origin of possible causal relations between subsystems. If we think we understand the "mecha-

States of Whole System	Determine States of	
	Subsystem A	Subsystem B
a	h	p
b	i	q
c	j	r
d	i	q
e	l	r
f	m	s
⋮	⋮	⋮

Figure 4.2. A is cause subsystem and B is effect subsystem.

States of Whole System	Determine States of	
	Subsystem A	Subsystem B
a	h	p
b	i	q
c	j	r
d	k	s
e	h	p
f	l	t
⋮	⋮	⋮

Figure 4.3.   Each subsystem is both cause and effect subsystem.

nism" that underlies the correlation, we express the relation in terms of "producing," etc. But even if we do not understand the "mechanism" or perceive any "mechanism" at all, a causal relation may still exist between systems as an empirical fact of correlation between their state spaces. Upon discovering a pair of causally related systems, we indeed do tend to search for a whole system of which the causally related systems are subsystems, in order to "understand" the causal relation. However, the existence of a causal relation and our knowledge of its existence in no way depend on our understanding or perceiving its underlying "mechanism."

And how does this abstract definition relate to our intuitive understanding of the concept of causal relation? First of all, I would like to warn against attaching too much importance to one's intuitions, at least in science. Intuitions are no more than thought habits. And since these habits developed as a result of limited experience, their appropriateness to phenomena lying outside that range of experience is suspect, at the least. (Need I remind you of the history of the theories of special relativity and quantum mechanics?) Second of all, in spite of that there is no contradiction between our abstract definition of causal relation and our intuitive understanding of the concept. We are used to the idea that, if a cause produces an effect, every time the cause is in the same state the effect is in the same state. Otherwise we would not have even considered the "cause" as being a cause to begin with.

For example, we are used to thinking that forces produce acceleration, that the forces are the cause and the acceleration is the effect, and not vice versa. And indeed, whenever the same set of forces acts on a given body, the same acceleration occurs. We cannot agree that the acceleration is the cause and the forces the effect, because the same acceleration may be correlated with different sets of forces (having the same resultant). Different sets of forces may be correlated with the same acceleration, while different accelerations are never correlated with the same set of forces. In abstract terms the system is the body, the forces acting on it, and its acceleration. The forces are subsystem *A* and the acceleration is subsystem *B*. The answer to the first question is affirmative; the same state of forces is always correlated with the same state of acceleration, and even different states of forces are correlated with the same state of acceleration. The answer to the second question is negative; the same state of acceleration is not always correlated with the same state of forces. Therefore the forces and the acceleration are in causal relation, with the forces the cause and the acceleration the effect.

Let us return to our example of applause. The system is the human body, all its actions, and all its noises. The hands are subsystem *A*, which is considered to have only two states: clapping and not clapping. Subsystem *B* consists of all body noises and is also considered to have only two states: clapping noise and not clapping noise (i.e., silence or any other noise). As we review all states of the whole system, we find a most astounding correlation between the states of *A* and the states of *B*: Whenever the hands clap, clapping noise sounds; and whenever clapping noise sounds, the hands are clapping. Thus the answers to both the first and the second questions are affirmative, and there exists a causal relation between hand clapping and clapping noise, where each may be taken as cause, as effect, or as both.

Another example is the spatiotemporal configuration of electric charge and current densities as the cause and the electric and magnetic field strengths as effect. Or we could have the scattering potential and incident wave function as cause and the scattered wave function as effect. The cause might be your present situation and recent history, with your present mood as effect. And so on.

Or consider the famous Rutherford experiment, in which alpha particles are scattered from stationary metal nuclei. If the exact locations of the target nuclei and the precise trajectory of each incident particle could be known, these would be the cause, and the effect would be the point where each scattered alpha particle hits the fluorescent detection screen. However, the locations of the target nuclei and the trajectory of each incident particle are not known. The best we can do is to confine the metal nuclei to a thin foil of known density and to know the statistical properties of a beam of alpha particles (particle flux, velocity distribution, etc.). Taking these as the cause, the effect certainly cannot be the final trajectory of a scattered particle or even its scattering angle, since scattering in all directions is actually observed in the experiment. The effect is, in fact, the statistical distribution of scattering angles for the scattered alpha particles, expressed by the angular-distribution density function or differential cross section.

Problems

1. Discuss a possible causal relation between day and night both in terms of empirical correlation and in terms of mechanism.

2. Consider a possible causal relation between thirst and drink.

3. Consider a possible causal relation between crime and punishment.

4. Consider a possible causal relation between life and death.

5. One hears the claim that the existence of some attribute or quality implies the existence of its opposite. For example, the existence of good implies the existence of evil. The philosophically inclined reader might consider this claim from the point of view of causal relation.

4.2.   THE EQUIVALENCE PRINCIPLE

The ultimate goal of our line of reasoning, which started with a discussion of the causal relation, is *the symmetry principle*. Along this line we must stop at two way stations, which cannot be avoided if we want not only to use the symmetry principle but to understand it. At the first station we must clarify to ourselves certain points concerning the nature of science. At the second stop we will show that the very existence of science implies a principle, which we call *the equivalence principle: Equivalent causes— equivalent effects*. These two issues are dealt with in the present section. In the following section the symmetry principle is derived as an almost immediate consequence of the equivalence principle, and we defer its formulation to there.

One of the cornerstones of science, as that concept is usually understood, is reproducibility: the possibility of repeating an experiment under the same conditions, but at a different time or place, and obtaining the same result. This possibility affords the objectivity that is essential for science to be a common, lasting endeavor rather than a set of private sciences or impossible altogether. We are not claiming that all the phenomena in the world fulfill the condition of reproducibility. For instance, the phenomena of parapsychology are notorious for their irreproducibility. This fact does not negate the possible existence of such phenomena nor does it necessarily invalidate their investigation. But as long as reproducibility is not achieved in parapsychology, the latter cannot be a subject of *scientific* research, as this is usually understood.

Another cornerstone of science is predictability: the possibility of predicting the result of as yet unperformed experiments. This means that, until proved otherwise, we labor under the assumption that human intelligence is capable of understanding nature sufficiently to allow the prediction of phenomena that have not yet been observed. Reproducibility alone, without predictability, does not make science. It only allows recording, cataloging, and classifying of experimental observations as public information, with no benefit beyond the compilation itself. Science begins to be possible only when order is perceived among the collected facts, and on the basis of this order the results of new experiments are predicted.

The scientific tool by which we predict the results of experiments is the theory. A theory is any conceptual recipe or mathematical formula or other means like these that, when fed data about the conditions of an experiment, gives us the experimental result to be expected. The first test of a new theory is performed by comparing it with past experiments and their results. If the results expected according to the theory match the results actually obtained, the theory advances to the next testing stage. If not, it goes back to the drawing board for corrections or overhaul or into the wastebasket. For the next test the theory must predict the results of as yet unperformed experiments. If it does that with continuing success, our confidence in it continually increases. However, a single failure is sufficient to invalidate a theory in spite of its past success.

Reproducibility and predictability are expressions or our realization that causal relations exist in nature. A scientific theory is the expression of a certain causal relation; the data that the theory receives represent the physical cause, and the results that the theory gives represent the physical effect. When a theory receives the same data, it always gives the same results, and that reflects the relation between the cause and the effect.

A scientific theory may be considered as including a sort of antianalog computer. We use the term "antianalog computer," since, whereas analog computers represent mathematical procedures by physical processes, scientific theories lead to representations of physical processes by (almost always) mathematical procedures. This antianalog computer is equipped with a set of terminals into which input may be fed and out of which output may be obtained. Different subsets of terminals might be used to feed and obtain different types of input and output. Some terminals might be exclusively either for input or for output, while others might serve for both input and output (though not during the same application). Every input uniquely determines an output. (It is possible, of course, that the same output might be the result of different inputs.) This gives us reproducibility and predictability. A physical theory must also supply a set of rules for translating a physical situation into input acceptable by the antianalog computer (and even for deciding whether a physical situation can be so translated) and for translating the output into physical terms.

As an example, consider classical vacuum electromagnetic theory with Maxwell's equations (James Clerk Maxwell, Scottish physicist, 1831–1879) and certain boundary conditions as its antianalog computer. The input might be functions, as a mathematical translation of a physical situation in terms of electric charge and current densities. The output is then functions, which are translated as electric and magnetic field strengths, giving the forces on test charges.

As another example, consider the theory of nonrelativistic quantum mechanics with the Schrödinger equation (Erwin Schrödinger, Austrian physicist, 1887–1961) and certain boundary conditions as its antianalog computer. Our input often consists of functions, as a mathematical translation of a physical situation in terms of potential and incident wave function. The output is then a function, which is translated as the scattered wave function and from which are derived probability, cross sections, and such measurable quantities.

Consider a physical system and a theory of its behavior. A system must have at least one pair of subsystems in causal relation for a theory of its behavior to be possible, since, as stated above, a theory is an expression of a causal relation in a system. Consider any such pair, a cause subsystem and an effect subsystem. Since the theory describes the behavior of the whole system, it must take the translation of the states of the cause subsystem as input to its antianalog computer, and the translation of the output of its antianalog computer must, by exhausting its total information content, determine states of the effect subsystem. Moreover, the state of the effect that is causally correlated with a given state of the cause must be consistent with the translation of the output that is uniquely produced by the antianalog computer as a result of receiving as input the translation of the given state of the cause. See Fig. 4.4.

States of causes and effects are actual physical situations occurring in actual physical systems. Each such state is unique; no two are identical. Although similar in many respects, a pair of states might still differ, for example, in their geographical locations, in their times of occurrence, or in the experimental apparatuses with which they are concerned. In fact, since no physical system is perfectly isolated from the rest of the uni-

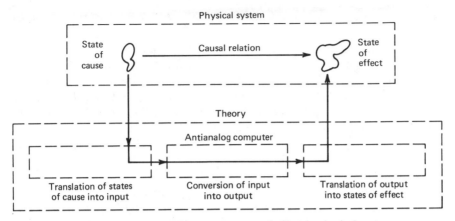

Figure 4.4. Relation of theory to cause and effect in physical system.

verse,* in order to make sure we are not missing anything we should, to be precise, take the whole universe as the only physical system worth investigating. Otherwise we run the danger of neglecting some influence, say from the Crab Nebula, that might be crucial. Then states of cause and effect subsystems would be characterized by an infinite number of parameters; we would have to take into consideration the state of every object, field, etc., in the universe, and at all times too, since these might possibly influence the states of subsystems.

This holistic point of view, with its emphasis on the uniqueness of situations, is unconducive to science. In its light everything appears so awesomely complicated that there seems to be no hope of understanding (in the sense of science) anything. Reproducibility and predictability seem utterly beyond reach. Science begins to be possible only when some order is discerned within the chaos, when similarities are remarked upon in spite of the differences. To do science one must, by decision, guess, or blissful ignorance, make assumptions about which influences are more or less important and on these assumptions analyze the world into quasi-isolated components. One must then investigate these relatively simple physical systems as if they were really isolated. The simplest systems must be investigated first. Then more complicated systems can be attacked by considering them as syntheses of simpler systems in interaction.

* Even if a physical system were isolated from the rest of the universe in all other respects, there would still remain that influence which, according to the Mach principle (Ernst Mach, Austrian physicist, psychologist, and philosopher, 1838–1916), is the origin of inertia. It would indeed be a great day if we were to succeed in abolishing inertia.

This very rough picture gives an idea of the attitude necessary for science to be possible.

Let me add that we have here, it seems to me, the main reason why science, and especially its predictability aspect, developed in the West and not in the East. While the dominant Eastern philosophies emphasized the oneness and wholeness of everything, making science extremely unlikely, if not altogether impossible, the Western *Weltansicht* encouraged analysis, making science most likely, if not inevitable. The different conceptions of the position of *Homo sapiens* vis-à-vis the rest of nature, tying in with religious considerations, were part of the whole scene. But let us return to our business.

We are thus led to the recognition that all scientific theories are doomed in principle to a sort of imprecision. The reasoning is that, if we took everything into consideration, we would be dealing solely with unique situations, reproducibility and predictability would be impossible, no order could be discerned within the chaos, and science would be inconceivable. It is only by ignoring certain aspects of physical reality that we can discover similarities, obtain reproducibility and predictability, discern order, and devise theories.

This essential ignoring is usually taken to be of two kinds. There is *ignoring in practice,* where we admit that there exists or might exist influence, but we assume it to be sufficiently weak that it is negligible compared with the effects under consideration. And there is *ignoring in principle,* where we assume that there is no influence at all. An example of ignoring in practice is the influence of the star Sirius on our laboratory experiments. We know there are gravitational and electromagnetic influences, but we ignore them, because we know they are negligible. We do not know whether there are other influences, but we assume that, if there are, they too are negligible. An example of ignoring in principle is the influence of "absolute" position on our experiments; we assume there is none. We assume the laws of nature are the same everywhere in the universe.

From the empirical point of view it is impossible to distinguish between precise lack of influence and sufficiently weak influence. Experimental investigations can only supply upper bounds for the strength of the effect. Indeed, it is possible that *in principle* there is no precise lack of influence and that all ignoring is ignoring in practice.* *In practice,* however, it is convenient to think in terms of the two kinds of ignoring.

Return now to a theory of behavior for a physical system in terms of

---

* Such matters and others are discussed in [M37].

antianalog computer and translation rules, as described previously. We have just convinced ourselves that the translation of states of a cause subsystem is essentially "weak" in the sense that certain aspects of the states are ignored in translation. Thus more than one state are translated into the same input to the antianalog computer. And similarly, the output does not contain sufficient information for unique determination of a state of the effect, so that more than one state of the effect is consistent with the translation of an output. This defines equivalent relations in the state spaces of the cause and effect subsystems, according to which states of the cause are equivalent if and only if they translate into the same input, and states of the effect are equivalent if and only if they are consistent with the translation of the same output.

For example, if a theory is consistent with the special theory of relativity, all states differing only by spatial or temporal displacements, rotations, or Lorentz transformations are equivalent. States that differ only in this way translate into the same input or are translated from the same output for a relativistic theory. In other words, a relativistic theory may take interest in any property of a state that is not a position, instant, orientation, or velocity connected with that state.

Or if a theory for a system of particles does not distinguish individual particles, all states differing only by permutation of particles are equivalent.

By this definition of equivalence in the state space of a cause subsystem all states belonging to the same equivalence subspace translate into the same input. This input uniquely determines an output according to the theory. And all states of the effect subsystem that are correlated with these states of the cause by the causal relation are translated from this output and are therefore equivalent with each other by definition. It is possible that also states of the effect that are correlated with states of the cause belonging to a different equivalence subspace, i.e., that translate into a different input, are equivalent with these, since different states of a cause may be causally related to the same state of an effect (but not vice versa) and since different inputs of a theory may determine the same output (while the same input always determines the same output). Figure 4.5 should help make this clear. Note in the example of the diagram that the state space of the cause decomposes into three equivalence subspaces, which translate into three different inputs. The antianalog computer converts two of these inputs into the same output and the third input into a different output. The two outputs translate into states of the effect, which decompose accordingly into two equivalence subspaces.

What we have just proved is *the equivalence principle*:

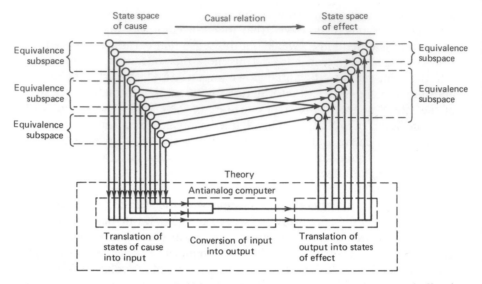

Figure 4.5. Theory imposes equivalence relations on state spaces of cause and effect in physical system, leading to the equivalence principle.

Equivalent causes—Equivalent effects

This formulation is relatively imprecise to avoid putting off those who have not yet studied its significance. The intention of the word "causes" is, of course, states of a cause, and the word "effects" similarly stands for states of an effect of the cause. The precise formulation is, therefore:

Equivalent states of a cause—
Equivalent states of its effect

And if we desire to put even more emphasis on the causal relation, we might insert an arrow:

Equivalent states of a cause →
Equivalent states of its effect

We should mention that the converse of the equivalence principle does not hold as a principle, since equivalent states of an effect are not necessarily causally correlated with equivalent states of a cause, as we saw.

We derived the equivalence principle as a direct consequence of the existence of science. Thus the existence of science, as this term is usually understood and as we presented it above, requires the existence of the equivalence principle, which is therefore a necessary condition for the existence of science. No equivalence principle (and thus no reproducibility) → no science!

The equivalence principle will be empty if and only if the equivalence relation a theory imposes on states of a cause is the trivial one, that every state is equivalent only to itself. Such a theory does not allow reproducibility, so it is not suitable for conventional science. However, such a situation could be envisaged for an all-encompassing theory of the universe (if one is possible), resulting, however improbably, from the all-inclusive contemplations of the holists or, more likely, from a synthesis of partial theories. Since there is only a single universe (at least we have access to only a single universe), reproducibility becomes meaningless. Thus it is the "weaknesses" or "impotencies" of scientific theories, their essential inability to distinguish among the members of certain sets of states, that make the equivalence principle nontrivial.

Let us summarize our line of reasoning that led to the equivalence principle. We started with the fact that experimentally reproducible phenomena exist in nature and that causal relations are found in nature. We assumed the validity of predictability, i.e., we assumed that the human intellect is capable of exploiting reproducibility to attain sufficient understanding of natural causal relations to enable prediction of as yet unobserved phenomena. The conceptual tool for the expression of causal relations in physical systems is the theory. Thus our assumption is that the human intellect is capable of inventing theories that describe the behavior of physical systems. This assumption seems to be well justified. A theory, by its essential "impotency," imposes an equivalence relation in a cause state space and in an effect state space: States of a cause are equivalent if and only if they translate into the same input, and states of an effect are equivalent if and only if they are translated from the same output. From this and from the character of any causal relation we obtained the equivalence principle:

$$\text{Equivalent states of a cause} \rightarrow$$
$$\text{Equivalent states of its effect}$$

Or in less precise language:

$$\text{Equivalent causes—Equivalent effects}$$

Problems

1. The cause is hearing a piece of music, and the effect is the amount of pleasure (or displeasure) experienced by the listener. The equiv-

alence relation of the cause is that pieces of music by the same composer are considered equivalent. Describe the causal relation that makes the equivalence principle hold.

2. The charges, positions, and velocities of source charges and test charges are the cause, and the accelerations of the test charges are the effect. Could states of the cause that are identical except that all velocities are reversed and all charges have opposite sign be equivalent? If so, what is the narrowest allowed equivalence relation of the effect?

3. Let the cause be the set of all forces acting on a body, and let the effect be the linear and angular accelerations of the body. Describe the most comprehensive equivalence relation of the cause for each of the following equivalence relations of the effect: (a) States with the same direction of linear acceleration are equivalent; (b) states with the same magnitude of linear acceleration are equivalent; (c) states with equal linear accelerations are equivalent; (d) states whose linear and angular accelerations are respectively equal are equivalent.

4. The system consists of crossed, uniform, electric and magnetic fields and a nonrelativistic charged particle moving in the fields, where the particle enters the fields moving perpendicularly to both. States of the particle are characterized by its mass, charge, and initial speed. For fixed electric field strength and given particle state the magnetic field strength can always be set so that the particle will pass through the crossed fields with no deflection. Let the state of the particle be the state of the cause and the magnetic field strength for zero deflection the effect. What is the most comprehensive equivalence relation of the cause?

5. The system is a nonrelativistic charged particle moving in a uniform magnetic field perpendicularly to the field. Such a particle moves in a circular orbit. States of the cause are characterized by the mass, charge, and speed of the particle, and strength of the magnetic field. The effect is the particle's orbit. What is the most comprehensive equivalence relation of the cause, if orbits of the same radius are considered equivalent?

6. It might happen that the phenomena of interest have to do with systems whose components are rather strongly interacting in general, so that their analysis into quasi-isolated, relatively simple subsystems is not at all straightforward or perhaps even impossible altogether. That might be the case for social, economic, and psychological phenomena, for example. Discuss reproducibility, predictability, order,

chaos, and the possibility of devising scientific theories in such cases. Consider parapsychological phenomena in this light.

4.3. THE SYMMETRY PRINCIPLE

Although the equivalence principle is fundamental to the application of symmetry in science, in practice it is another principle, called *the symmetry principle* and derived directly from the equivalence principle, that is usually used:

> *The symmetry group of the cause is a subgroup*
> *of the symmetry group of the effect*

Or less precisely:

> *The effect is at least as symmetric as the cause*

See [A17].

To prove the symmetry principle we first define the terms used in it. Consider again a physical system and a theory of its behavior. Any state u of the system implies a state for each of its subsystems and implies especially state u_c for a given cause subsystem and state u_e for a given effect subsystem. The theory imposes an equivalence relation for the state space of the cause subsystem (giving the "equivalent causes" in the equivalence principle) and an equivalence relation for the state space of the effect subsystem (giving the "equivalent effects" in the equivalence principle), as discussed in the preceding section.

Define cause-equivalence, denoted $\overset{c}{\equiv}$, for the state space of the whole system as follows: Two states of the whole system are cause-equivalent if and only if the states of the cause subsystem implied by them are equivalent. Symbolically,

$$u_c \equiv v_c \Leftrightarrow u \overset{c}{\equiv} v$$

for states u and v of the whole system. Similarly define effect-equivalence, denoted $\overset{e}{\equiv}$, for states of the whole system: Two states of the whole system are effect-equivalent if and only if the states of the effect subsystem implied by them are equivalent.

$$u_e \equiv v_e \Leftrightarrow u \overset{e}{\equiv} v$$

for states u and v of the whole system. See Fig. 4.6.

The equivalence principle states that if the states of the cause subsystem implied by states of the whole system are equivalent, then the states of the effect subsystem implied by these same states of the whole system are also equivalent. Symbolically,

$$u_c \equiv v_c \Rightarrow u_e \equiv v_e$$

for states u and v of the whole system. From our definitions we conclude that cause-equivalence implies effect-equivalence for the state space of the whole system. Or, symbolically,

$$u \overset{c}{\equiv} v \Rightarrow u \overset{e}{\equiv} v$$

for states u and v of the whole system. See Fig. 4.7.

We now define the symmetry group of the cause as the symmetry group of the whole system for cause-equivalence in its state space. Similarly, the symmetry group of the effect is defined as the symmetry group of the whole system for effect-equivalence. Since cause-equivalence implies effect-equivalence, it follows that every element of the symmetry group of the cause must necessarily also be an element of the symmetry

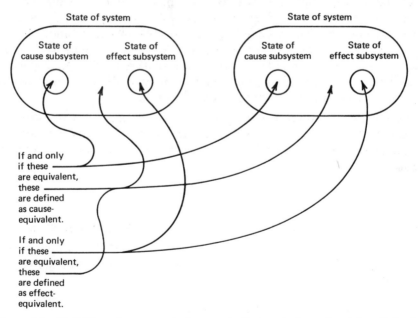

Figure 4.6. Definition of cause-equivalence and effect-equivalence for states of physical system.

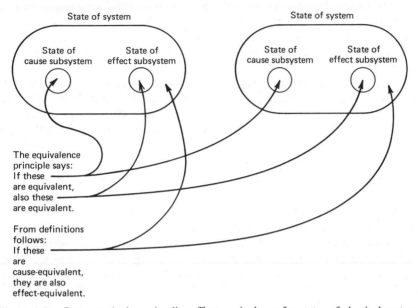

Figure 4.7. Cause-equivalence implies effect-equivalence for states of physical system.

group of the effect. There might, of course, be symmetry transformations of the effect that are not also symmetry transformations of the cause. This proves the symmetry principle, that the symmetry group of the cause is a subgroup of the symmetry group of the effect. And referring to our discussion of quantification of symmetry in Section 3.6, we obtain the alternative, less precise formulation of the symmetry principle, that the symmetry of the effect is at least that of the cause.

4.4. MINIMALISTIC USE OF THE SYMMETRY PRINCIPLE

The symmetry principle can be used to set a lower bound on the symmetry of the effect or to set an upper bound on the symmetry of the cause. The former use, which we call the minimalistic use, is characteristic of most practical, technological, and textbook problems, where the theory and cause are known for a given system and it is desired to find some or all of the effect. If the cause is known, its symmetry can be worked out. The symmetry principle then states that the effect also possesses this symmetry (and possibly more). The knowledge of the minimal symmetry of the effect is often sufficient for solving the problem fully or partially or at least for simplifying it to some degree. The minimalistic use of the symmetry principle is exploited in science in widely varying degrees of sophistication. Here we offer some examples of the more simple-minded applications, in order to keep the symmetry considerations in the foreground.

In our first example we find the direction of acceleration of a resting test charge due to a spherically symmetric charge distribution. What we are doing, of course, is finding the direction of the electric field of a spherically symmetric charge distribution, but we prefer to formulate the problem in spatiotemporal terms, in terms of acceleration, rather than in terms of the electric field, since the transformation properties of the electromagnetic field are not trivial, and we prefer not getting involved with them here. Note that we also avoid formulating the problem in terms of force, since, although no harm would be done, we still feel we are standing on firmer ground when we assign transformation properties to acceleration rather than to force. Also note that this example is equally applicable to any central force field, as no use is made of the Coulomb force law (Charles Augustin de Coulomb, French physicist, 1736–1806) specifically.

The cause in this example consists of the charge distribution and the test charge. The effect is the acceleration of the test charge. Refer to Fig. 4.8. The symmetry group of the cause is the group of all rotations about the axis passing through the test charge and the center of the charge distribution and reflections through all planes that contain this axis. By the symmetry principle the effect, the acceleration, must also have this symmetry at least. Since any nonradial component the acceleration might have would not be carried into itself by all the symmetry transformations of the cause, the acceleration must therefore be radial. See [A49].

In the next example we find the direction of acceleration of a test charge moving parallel to a straight, infinitely long, current-carrying wire. Although the force on the test charge is magnetic, this fact will not enter our discussion. And we purposely formulated the problem in terms of the acceleration of a test particle, rather than in terms of the magnetic field around the wire, in order to avoid the complicating side issue of the

Figure 4.8. System consisting of spherically symmetric charge distribution and resting test charge, as cause, and acceleration ($\vec{a}$) of test charge, as effect.

Figure 4.9. System consisting of infinite straight current and test charge moving parallel to it, as cause, and acceleration ($\vec{a}$) of test charge, as effect.

transformation properties of the magnetic field under reflection, a common source of trouble to the unwary.

The cause consists of the current and the moving test charge. The effect is the acceleration of the test charge. Refer to Fig. 4.9. (Although in the figure the current was given the same sense as the velocity of the test charge, this point is immaterial, and the opposite sense would serve as well.) The symmetry group of the cause contains the transformation of reflection through the plane containing the wire and the test charge (the plane of the paper in the figure). Since by the symmetry principle this must also be a symmetry transformation of the effect, the acceleration can have no component perpendicular to this plane.

The symmetry group of the cause also contains the more complicated transformation consisting of the consecutive application of temporal inversion and reflection through the plane perpendicular to the current and containing the test charge (at any instant). The temporal-inversion transformation by itself reverses the senses of all velocities, in our example both that of the test charge and that of the moving charges forming the current. Since both the current and the velocity of the test charge are perpendicular to the reflection plane of the second part of the transformation, their senses are reversed again by the reflection transformation. Thus the combined transformation indeed leaves the cause intact. See Fig. 4.10. But how does it affect the effect, the acceleration, of which it must be a symmetry transformation? Temporal inversion by itself does not change acceleration, since the dimension of acceleration is even-powered in time, (length) $\times$ (time)$^{-2}$. And the reflection reverses the component of acceleration perpendicular to the reflection plane. See Fig. 4.10. So symmetry of the effect further limits the possible directions of the acceleration and leaves it perpendicular to the velocity of the test particle and pointing either toward or away from the wire.

In our next example we prove that the orbit of a planet about its sun lies completely in a plane and that this plane passes through the center of the sun. Our only assumption is that the sun and planet are each

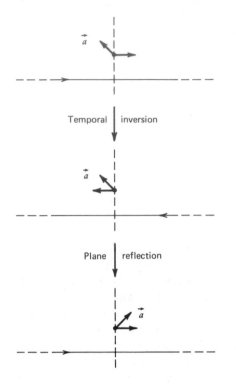

Temporal | inversion

Plane | reflection

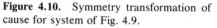

Figure 4.10. Symmetry transformation of cause for system of Fig. 4.9.

spherically symmetric. We assume nothing about the nature of the force between the two. (The reasoning will be valid also for a spherically symmetric body in any central force field.)

Consider the situation at any instant. The planet has a certain position and a certain instantaneous velocity relative to the sun. Of all the planes containing the line connecting the centers of the planet and the sun only one is parallel to the direction of the planet's instantaneous velocity. We call this the plane of symmetry. See Fig. 4.11. The cause consists of the sun and the moving planet. The effect is the planet's acceleration.

The cause has reflection symmetry with respect to the plane of symmetry: The sun and planet are each reflection symmetric with respect to any plane containing their centers, since they are spherically symmetric, and the plane of symmetry contains both centers. The position of the planet is not changed by reflection through the plane of symmetry, again since the plane contains the planet's center. The direction of the planet's instantaneous velocity is parallel to the plane of symmetry, so this velocity is also invariant under reflection through the plane. Thus the cause is reflection symmetric with respect to the plane of symmetry. The effect, the acceleration of the planet, must then have this symmetry at least. Therefore, the direction of the planet's acceleration must be parallel to the plane of symmetry. (If the acceleration had a component perpendicular to the plane, it and its reflection image would not coincide.) So the planet's velocity, which is parallel to the plane of symmetry, undergoes a change of velocity parallel to the plane of symmetry and thus remains parallel to this plane. In this way we see that the planes of symmetry of the system at all instants are in fact one and the same plane and that the motion of the planet is confined to this plane.

Our next example of application of the symmetry principle concerns electric currents. Consider the DC circuit of Fig. 4.12, probably good for nothing but an exercise. The six emf (voltage) sources, the 11 resistors, and their connections are the cause. The effect consists of the resulting currents in the various branches of the circuit and the potential differences between all pairs of points of the circuit. By Kirchhoff's first law (Gustav Robert Kirchhoff, German physicist, 1824–1887) we arbitrarily designate seven of the currents i_1, i_2, i_3, i_4, i_5, i_6, i_7 and express the other six currents in terms of these as in the figure. Kirchhoff's first and second laws then give us a set of seven simultaneous linear equations for the seven unknown currents, which we do not show here. These equations can be solved and the solutions used to calculate the other six currents. With all the currents known the potential difference between any pair of points can be calculated. For example, referring to the figure, we might want to know the potential difference between any pair of the points marked A, B, C, D, E, F, such as V_{AB}, V_{AD}, V_{EF}.

A glance at Fig. 4.12 reveals that the cause has symmetry, twofold rotation symmetry about the axis through the 2-kΩ resistor and perpendicular to the plane of the paper. So the symmetry principle can be

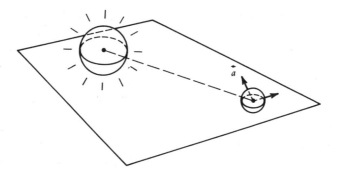

Figure 4.11. System consisting of sun and moving planet, as cause, and acceleration ($\vec{a}$) of planet, as effect. Plane of symmetry is indicated.

Figure 4.12. DC circuit.

invoked, and the effect, the currents and potential differences, must be at least as symmetric as that. To help us see what that gives we rotate the system by 180°, as in Fig. 4.13. Now compare the rotated system of Fig. 4.13 with the unrotated system of Fig. 4.12. Symmetry of the effect gives us for the currents $i_1 = i_4$, $i_2 = i_5$, $i_3 = i_6$, $i_7 = -i_7$, so that $i_7 = 0$. For the potential differences symmetry of the effect gives $V_{AD} = V_{DA}$, $V_{BE} = V_{EB}$, $V_{CF} = V_{FC}$. And since by the nature of potential differences $V_{DA} = -V_{AD}$, etc., we have $V_{AD} = V_{DE} = V_{CF} = 0$. Symmetry of the effect also gives us relations among the potential differences, such as $V_{AF} = V_{DC}$, $V_{CE} = V_{FB}$, etc.

Thus the symmetry principle gives us more than half the solution to the problem in this example. Instead of solving seven simultaneous equations for seven unknowns, we need to solve only three equations for three currents. Three of the desired potential differences are now known, and only half of the others need to be calculated.

Our next example is again an electrical one. In this example the system has a higher degree of symmetry than in the preceding one, and the symmetry principle makes the complete solution quite simple, although the explaining of it is a bit drawn out. We solve the well-known problem of finding the resistance of a network of 12 equal resistors connected so that each resistor lies along one edge of a cube, where the resistance is measured between diagonally opposite vertices of the cube, such as be-

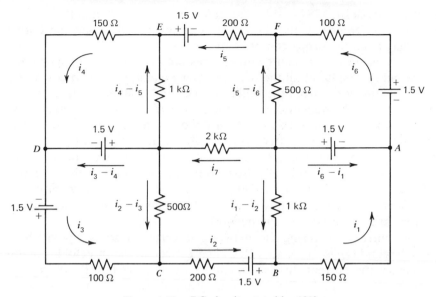

Figure 4.13. DC circuit rotated by 180°.

tween vertices A and H in Fig. 4.14. Let r denote the resistance of each resistor. We imagine a potential difference V applied between vertices A and H. As a result current i enters the network at A, branches through the various resistors, and leaves the network at H. The resistance of the network from A to H is then $R = V/i$ by Ohm's law (Georg Simon Ohm,

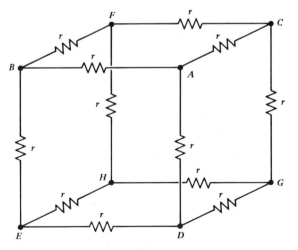

Figure 4.14. Cube of resistors.

German physicist, 1787–1854). The cause is the network and the applied potential difference, while the currents in the resistors and the corresponding potential drops comprise the effect.

The cause in the present example does not possess the full symmetry of the cube, in spite of all the resistors being equal, since vertices A and H are distinguished from the other vertices and from each other, as the current enters the network at A and leaves at H. So the symmetry transformations of the cause consist of only those symmetry transformations of the cube that do not affect vertices A and H: rotations by 120° and 240° about the diagonal AH (i.e., diagonal AH is an axis of threefold rotation symmetry) and reflections through each of the three planes $ABHG$, $ACHE$, and $ADHF$ as in Fig. 4.15. Then by the symmetry principle the effect must also have this symmetry. We can use this fact to find the current in each resistor of the network.

The current i entering the network at vertex A splits among three branches and flows to vertices B, C, and D. Due to the threefold rotation symmetry the current divides equally, so that current $\frac{1}{3}i$ flows in each branch AB, AC, and AD as in the diagram of Fig. 4.16. The current $\frac{1}{3}i$ entering vertex B then divides again between two branches and flows to vertices E and F. It divides equally between the two branches because

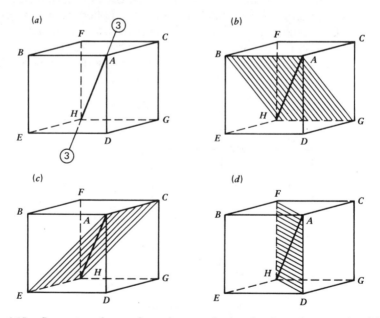

Figure 4.15. Symmetry of cause for resistance of cube of resistors between A and H. (a) Axis of threefold rotation symmetry. (b)–(d) Planes of reflection symmetry.

Figure 4.16. Rotation symmetry requires equal first division of current entering network.

of the reflection symmetry with respect to plane *ABHG*. Thus current $\frac{1}{2}(\frac{1}{3}i) = \frac{1}{6}i$ flows in each branch *BE* and *BF*. Similar reasoning also gives current $\frac{1}{6}i$ in each of the branches *CF*, *CG*, *DG*, and *DE* as in Fig. 4.17.

After that stage the symmetry takes care of itself. Current $\frac{1}{6}i$ enters vertex *E* from each of vertices *B* and *D*, producing current $2 \times \frac{1}{6}i = \frac{1}{3}i$ leaving *E* and flowing to *H*. Similarly, current $\frac{1}{3}i$ enters *H* from each

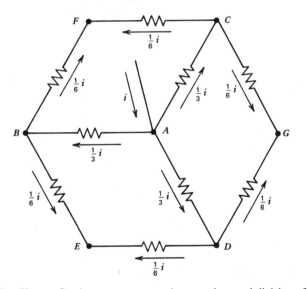

Figure 4.17. Plane-reflection symmetry requires equal second division of current.

of vertices F and G. The three currents $\frac{1}{3}i$ entering vertex H join to give current $3 \times \frac{1}{3}i = i$ leaving the network. That is illustrated in Fig. 4.18.

Now the potential difference V between A and H equals the sum of potential drops between those two vertices, where the sum may be calculated over any continuous path connecting A and H. Let us take path $ABEH$. By Ohm's law the potential drop on a resistor equals the product of the current in the resistor and its resistance. So, referring to Fig. 4.18, the potential drop from A to B is $\frac{1}{3}ir$. From B to E the potential drop is $\frac{1}{6}ir$, and from E to H it is $\frac{1}{3}ir$. Adding these together, we obtain $V = \frac{5}{6}ir$. Thus the resistance of the network between A and H is

$$R = V/i = \tfrac{5}{6}ir/i = \tfrac{5}{6}r.$$

This is the solution of the problem.

In the last two examples note that it is not the actual geometry of the circuits that is important but rather their electric structure. Thus it is not the geometric symmetry of the circuits that really interests us but rather the permutation symmetry of their components. It is true that we expressed that permutation symmetry in geometric terms in order to keep our considerations as familiar as possible; however, the former circuit does not have to be laid out as nicely as in Fig. 4.12 nor does the latter network actually have to lie along the edges of a cube for our symmetry arguments to hold. Thus it is really permutation symmetry that we are using, based on the functional, rather than geometric, equivalence of parts of the circuit. This point is important to keep in mind in the application of symmetry to physical systems: Functional equivalence of parts of a system can be a source of permutation symmetry.

Another point to note in the last two examples is that each problem has a unique solution. So naturally the solution was chosen to be the effect in each case, and application of the symmetry principle was straightforward. Uniqueness of solution is characteristic of passive electric circuits. That is not always the case, however, when circuits contain active components (transistors, vacuum tubes, etc.). Then more than one solution might be possible. What is the effect in this case and how is the symmetry principle applied? Read on.

Our next example of application of the symmetry principle is of somewhat different character. It is often assumed that the electric charge distribution on the surface of an isolated charged conducting sphere is homogeneous, i.e., that the surface charge density in such a case possesses spherical symmetry. And if any explanation is offered at all, it is invariably simply a reference to "symmetry considerations." Now it is indeed true that the surface charge density of an isolated charged con-

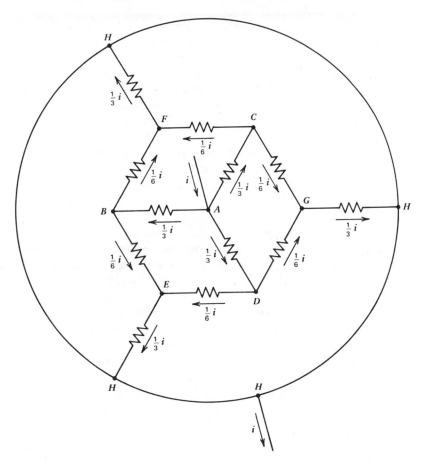

Figure 4.18. Rejoining of divided currents in two stages before leaving network.

ducting sphere possesses spherical symmetry. That can be shown, using the fact that the surface of a conducting sphere is an equipotential surface and relying on uniqueness considerations. That has not been shown, as far as I know, by proving that the spherically symmetric configuration is the configuration of lowest electrostatic potential energy. As for "symmetry considerations," they certainly do not require a spherically symmetric charge distribution, as we will see, although they do allow it.

How does the sphere get its charge to begin with? We must charge it. We then remove the charging device far enough away so that the sphere can be considered isolated. The cause consists of the charging device and the sphere, and the effect is the final charge distribution on the sphere. The most symmetric practically attainable cause I can think of is attained by keeping the charge source very far from the sphere at all times and connecting the two through a straight wire that touches the sphere perpendicularly to the sphere's surface, so that the line of the wire passes through the center of the sphere. When charging is completed, the wire is withdrawn from the sphere along its line. (Alternatively, a spherical charge source could be put in contact with the sphere and then removed along the line containing the two centers.) Thus the most symmetry we can give the cause is axial and reflection symmetry, symmetry under the group of all rotations about the line of the wire (or the line containing the centers) and reflections through all planes containing this line. See Fig. 4.19. This is not spherical symmetry, which is symmetry under the group of all rotations about all axes through the center of the sphere and reflections through all planes containing the center. The latter group obviously includes the former as a proper subgroup. Indeed, the sphere itself is spherically symmetric. However, the *total* cause is only axially and reflection symmetric. So the only symmetry we can demand of the effect, the charge distribution, is also axial and reflection symmetry, although spherical symmetry is not excluded by the symmetry principle.

If we do not know or prefer to ignore the way the sphere is charged and wish to consider the sphere itself as the cause, the effect will certainly not be the charge distribution, which might depend on the way the sphere is charged, but the set of all allowed charge distributions. Since the cause in this case is spherically symmetric, so is the effect, by the symmetry principle. Thus, even though no single allowed charge distribution has to be spherically symmetric, given any one allowed charge distribution, all those obtained from it by any rotation about any axis through the center of the sphere or reflection through any plane containing the center are also allowed charge distributions. See [A49].

This brings us to the following warning. If the state of what is chosen as the cause in a problem is not sufficient to determine a unique solution to the problem, then the effect cannot be the solution, but is rather the family of all solutions consistent with a state of the cause. As another example of this, given as the cause an equation possessing some symmetry, the effect is not any solution of the equation, and indeed no solution

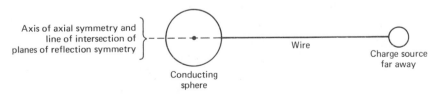

Figure 4.19.   Charging of conducting sphere.

need possess the full symmetry of the equation. The effect is the family of all solutions of the equation, and that is what must have the symmetry of the equation. Given any one solution, the application to it of any symmetry transformation of the equation produces another solution (or even possibly the same solution). If, however, additional conditions are involved, such as boundary or initial conditions or constraints, they may be included in the cause and will usually reduce its symmetry.

To give a more specific example, consider the algebraic equation

$$x^2 - 4 = 0$$

or any other polynomial equation involving only either even or odd powers of the unknown variable. Every such equation is symmetric under the transformation

$$x \rightarrow -x.$$

(The odd polynomials change sign under this transformation, but the equations are the same.) If each root of such an equation had the same symmetry, the only allowed root would be $x = 0$. However, it is the set of all roots, and not each root, of the equation that must possess the symmetry of the equation, with the result that for equations of the type we are discussing all nonzero roots must come in positive–negative pairs. For instance, the roots of the equation just presented are $x = \pm 2$.

Or consider the harmonic-oscillator differential equation

$$\frac{d^2y}{dt^2} + a^2y = 0, \qquad a = \text{const.}$$

It possesses symmetry under each of the following transformations:

  temporal displacement $t \rightarrow t + b$

for arbitrary constant b,

  temporal inversion $t \rightarrow -t,$

  spatial dilatation $y \rightarrow cy$

for arbitrary, positive constant c, and

  spatial inversion $y \rightarrow -y.$

Since no initial conditions are specified, the family of solutions of this equation must have the same symmetry as the equation. Thus, if

$$y = f(t)$$

is a solution, so must also

$$y = f(t + b),$$
$$y = f(-t),$$
$$y = cf(t),$$
$$y = -f(t)$$

be solutions for arbitrary allowed constants b and c. For the oscillator equation the family of all solutions can be written in the form

$$y = A \sin(at + B) = C \sin at + D \cos at$$

for all constant A and B or C and D. Under the temporal-displacement transformation the general solution becomes

$$y = A \sin(at + ab + B)$$
$$= A \sin(at + B'),$$
$$B' = ab + B.$$

Since B' ranges over the same values as B, we have recovered the same family. Under the temporal-inversion transformation the general solution becomes

$$y = A \sin(-at + B)$$
$$= -A \sin(at - B)$$
$$= A' \sin(at + B'),$$
$$A' = -A,$$
$$B' = -B,$$

and again we find that the family is symmetric. Under the spatial-dilatation transformation the general solution becomes

$$y = cA \sin(at + B)$$
$$= A' \sin(at + B),$$
$$A' = cA;$$

symmetry again. And under the spatial-inversion transformation the general solution becomes

$$y = -A \sin(at + B)$$
$$= A' \sin(at + B),$$
$$A' = -A,$$

and is symmetric. If initial conditions are imposed by specifying the values of the solution and its derivative for some value of t, the special solution will in general not possess any of the above symmetries, although in certain cases it might retain one or more of them.

Note that each special solution of the oscillator equation has the property of periodicity, i.e., it is symmetric under a certain minimal temporal displacement and integer multiples of it,

$$y\left(t + \frac{2\pi}{a} n\right) = y(t)$$

for

$$n = 0, \pm 1, \pm 2, \ldots .$$

(This symmetry group is a subgroup of the temporal-displacement group.) This symmetry does not follow from the symmetry of the equation but can be discovered only by actually solving the equation. In comparison, the equation

$$\frac{d^2 y}{dt^2} - a^2 y = 0,$$

similar to the oscillator equation, possesses the same four symmetries that we found for the oscillator equation but does not have periodic solutions. In fact, its general solution can be written

$$y = C \sinh at + D \cosh at,$$

and the hyperbolic functions are not periodic. This family of solutions must and does possess the four symmetries of its equation.

Problems

1. Extend the discussion of the spherically-symmetric–charge-distribution example, the first example of this section, to show that for

given charge distribution the only geometric variable that the magnitude of the test charge's acceleration can depend on is the distance between the test charge and the center of the distribution.

2. For the example of a current-carrying wire and moving test charge extend the argument to show that the only geometric variable on which the magnitude of the test charge's acceleration can depend is the distance between the test charge and the wire.

3. For a uniformly charged infinite plane prove by the symmetry or equivalence principle that: (a) the direction of the electric field produced by such a charge distribution is perpendicular to the plane; (b) the only geometric variable that the field strength can depend on is the distance from the plane; and (c) at equal distances on either side of the plane the field strengths have equal magnitudes and opposite senses.

4. In the sun-and-planet example of this section the plane of symmetry is not uniquely defined when the planet is at rest or moving directly toward or away from the sun. Use the symmetry principle to prove that in these cases the orbit of the planet is on a straight line through the center of the sun.

5. Using the symmetry principle, show for the cube of resistors discussed in this section that vertices B, C, and D are equipotential points and so are vertices E, F, and G. Prove this also by electrical considerations in the diagram of Fig. 4.18.

6. Use the symmetry principle to find the resistance of a network of 12 equal resistors connected to form the edges of an octahedron, where the resistance is measured between a pair of opposite vertices.

7. According to Hermann Weyl (German-American mathematician, 1885–1955) [S24], Ernst Mach "tells of the intellectual shock he received when he learned as a boy that a magnetic needle is deflected in a certain sense, to the left or to the right, if suspended parallel to a wire through which an electric current is sent in a definite direction." See Fig. 4.20. Explain why this phenomenon seems to violate the symmetry principle, to young Mach's shock. Explain why it really does not.

8. Find a useful or interesting nontrivial problem that can be completely solved or at least greatly simplified by the symmetry principle.

9. Find the symmetries of each of the following equations and, solving

Figure 4.20. Mach's dilemma.

the equation, show that the set of roots of each equation possesses the same symmetries:

(a) $x^2 + 4 = 0$,

(b) $x^5 - x^3 - 6x = 0$,

(c) $\cos x = \sqrt{(1 - \sin^4 x)}$.

10. Find the symmetries of each of the following differential equations and show, by solving it, that the general solution of each equation possesses the same symmetries. Then impose the initial condition $y(0) = 1$ on each solution and check which of the symmetries are lost and which are retained.

(a) $\dfrac{d^2y}{dt^2} = a,\ a = \text{const}$;

(b) $\dfrac{dy}{dt} = ay,\ a = \text{const}$;

(c) $y\dfrac{dy}{dt} = a,\ a = \text{const}$.

11. Use the symmetry principle to show that the complex roots of a polynomial equation with real coefficients must be in conjugate pairs. What is the symmetry transformation involved here?

4.5.   MAXIMALISTIC USE OF THE SYMMETRY PRINCIPLE

The problems of basic scientific research are usually opposite to those for which the symmetry principle is used minimalistically. In basic research it is the effect that is given, and we try to find the cause. The effect in such a problem is one or more experimental phenomena, and we attempt to find a theory explaining these phenomena as being brought about by some cause. A theory is considered to be better the more different phenomena it explains and the simpler the cause that is supposed to be producing them. Although simplicity is not a standardized concept and is largely a matter of taste, it is generally agreed that symmetry contributes greatly to simplicity. So in devising a theory for a given set of experimental phenomena, we usually assume as symmetric a cause as possible. And how symmetric can a cause be? Here the symmetry principle serves to set an upper bound on the symmetry of the cause; the cause can be no more symmetric than the effect. This is what we call the maximalistic use of the symmetry principle.

So we must first identity the symmetry of the phenomena we wish to explain. (This symmetry is often far from obvious.) We then construct our theory so that the cause will have just the same symmetry, if possible. If it is not possible to assign maximal symmetry to the cause, we must assume a less symmetric cause and include in the theory an explanation of why the effect is more symmetric than the cause. But we may never assume that the cause has a higher degree of symmetry than the phenomena being explained. This would violate the symmetry principle and thus also the equivalence principle.

What most often happens, though, is that the symmetry of a set of phenomena is only approximate. (We discuss approximate symmetry in Chapter 5; for the present let us make do with whatever idea about it, however vague, you might have.) In such a case the first step toward a theory is to determine the ideal symmetry that is only approximated by the phenomena. This can be very difficult, if the symmetry is far from exact. Then to obtain as symmetric a cause as possible we try to construct a theory such that the cause will have a dominant part (in an appropriate sense), possessing the ideal symmetry of the effect, and another, symmetry-breaking part, which does not have that symmetry. In the (possibly hypothetical) limit of complete absence of symmetry breaking, the dominant part of the cause produces the ideal symmetry of the phenomena, while the symmetry-breaking part brings about the deviation from ideal symmetry. A complete theory will contain the symmetry-breaking mechanism within its framework. But sometimes the cause of the breaking must be left as a mystery to be cleared up when more experimental facts are known or a better theory can be found.

For an example of the maximalistic use of the symmetry principle we turn to nuclear physics. The basic problem of nuclear physics is the strong interaction, the force that binds protons and neutrons together to form nuclei. This nuclear force is not yet completely understood. On the other hand, the electromagnetic interaction among protons and neutrons as well as the Pauli exclusion principle (Wolfgang Pauli, Swiss–Austrian physicist, 1900–1958), stating in the present context that no two protons or no two neutrons can be in the same quantum state, are very well understood. The effect consists of such phenomena as the various properties of all kinds of nuclei and the results of scattering experiments, in which protons and neutrons are made to collide. The effect is found to exhibit the following approximate symmetry. Two kinds of nuclei differing only in that one of the neutrons in one kind is replaced by a proton in the other have certain similar properties (though their electric charges, for example, are clearly different); in scattering experiments similar results are obtained whether the interacting particles are two protons, two neutrons, or a proton and a neutron.

Thus nuclear phenomena are approximately symmetric under interchange of proton and neutron. This symmetry is called charge symmetry, since the major difference between the proton and the neutron is their different electric charges. The symmetry principle then suggests that we assume that the nuclear force is exactly charge symmetric, is completely blind to any difference between the proton and the neutron. The symmetry-breaking factors are assumed to be the electromagnetic interaction, which discriminates between proton (electrically charged) and neutron (electrically neutral), and the Pauli principle, which discriminates between identical particles and different particles. (In the hypothetical limit of no electromagnetic interaction and no Pauli principle, when the symmetry-breaking factors are absent, the approximate charge symmetry of nuclear phenomena should become exact.) This assumption has, in fact, proved to be very successful.

Another example of the maximalistic use of the symmetry principle is in the study of elementary particles and their interactions (of which nuclear physics is actually just a special case). Here the symmetries of the experimental phenomena are much more complicated and much more approximate (i.e., less nearly exact, except for certain exact symmetries) than in nuclear physics, and the theorists are searching for ways to take them into account in their theories.

4.6. QUANTUM SYSTEMS

This section is intended for readers with a good understanding of quantum theory and its Hilbert-space formulation. If this description does not fit, you are advised to skip the section.

The application of symmetry in quantum systems is straightforward and successful. It is, in fact, one of the greater success stories of theoretical science. This is due to the special nature of quantum state spaces, specifically to their being linear vector spaces, which makes it possible to work with realizations of transformation and symmetry groups by groups of matrices acting on the components of the elements (vectors) of quantum state spaces. Such realizations are called linear representations, or representations. The detailed study of the application of symmetry in quantum systems, which is beyond the scope of this book, is well presented in a considerable number of more advanced texts.

It is worthwhile, however, to devote a brief discussion to causes and effects in quantum systems. Causes and effects, as defined previously, are subsystems of the systems under consideration, and subsystems of quantum systems are themselves quantum systems. Thus causes and effects in quantum systems are quantum systems with all the consequences thereof. For example, in a scattering experiment, if the sharply peaked momentum distribution in the incident beam is part of the cause, one may not expect that the positions of the individual particles in the beam might also be included in the cause, since the latter are in principle indeterminate in the given setup. As for the effect, neither particle trajectories nor scattering angles of individual particles can be part of it. The effect is, in fact, the wave function, especially the "scattered" wave function, from which probabilities and differential cross sections can be calculated.

If that sounds like the Rutherford experiment, you are quite right, and we should compare the classical and quantum theories of such scattering with regard to cause and effect in each kind of theory. In the classical approach the cause could in principle be the initial position, velocity, and orientation of each incident particle and the position and orientation of each target particle. The effect would then be the point where each scattered particle hits the fluorescent detection screen and causes it to flash. However, such a cause is not practically realizable (and is, of course, precluded by quantum principles), and we must make do instead with the statistical properties of the incident beam and the target material. The effect is then correspondingly reduced to statistical properties of the scattered particles. In nonrelativistic quantum mechanics, on the other hand, the cause is the incident wave function and the scattering potential, which indeed involve statistical properties of the incident and target particles but are different from the classical cause. For instance, the incident wave function has a nonclassical phase (over which we admittedly have no control, but which could be of importance in its relation to other phases in the system). The effect is the scattered wave function, which also contains a phase in addition to statistical information about the scattered particles.

The vacuum state of quantum theory is also due for some discussion.

Invariably, even in the most fundamental theories, the properties of the vacuum state are part of the assumptions of the theory, rather than derived from other assumptions. Thus the vacuum state is part of the cause. This must be kept in mind, especially when the vacuum state is assumed to possess a lower degree of symmetry than the rest of the cause. It is the symmetry of the total cause, including the vacuum state, that, by the symmetry principle, must appear in the effect. Since the vacuum state is crucial in determining the properties of the physical states, as the effect, it should come as no surprise that the physical states are not as symmetric as the cause without the vacuum state, and the situation is in no way a violation of the symmetry principle.

I brought up that point because it has often been presented as an apparent violation of the symmetry principle. The symmetry principle has nothing to fear from such or other *apparent* violations. It has be contrived so that, as long as we are concerned with conventional science (reproducibility, predictability, theories, etc.), it is inviolable.

Approximate Symmetry and Spontaneous Symmetry Breaking

In this chapter we briefly discuss approximate symmetry, based on the concept of a metric for a state space of a system. We present the concepts of approximate symmetry transformation, approximate symmetry group, goodness of approximation, exact symmetry limit, broken symmetry, and symmetry-breaking factor. We then more substantially discuss spontaneous symmetry breaking, situations where the symmetry principle—that the effect is at least as symmetric as the cause—*seems* to be invalid. The essence of the matter is the stability of the effect under perturbations of the symmetry of the cause.

The previous chapters of this book are prerequisite for this chapter. No additional mathematical knowledge is needed. The scientific background of the preceding chapter should more than suffice. The paragraph concerned with quantum theory at the end of the chapter is intended, of course, for the initiated.

5.1. APPROXIMATE SYMMETRY

To be able to discuss approximate symmetry we need the concept of *approximate symmetry transformation*, any transformation that maps every state to an image state that is "nearly" equivalent to the object state. And just what does "nearly" equivalent mean? For that we must soften the all-or-nothing character of the equivalence relation, upon which symmetry is based, in order to allow, in addition to equivalence, varying degrees of inequivalence. The way to do that is to define a *metric* for a state space of a system, a "distance" between every pair of states, such that zero "distance" indicates equivalence and positive "distances" represent degrees of inequivalence.

More precisely, a *metric* is a bivariable, nonnegative, real function on state space $d(\quad,\quad)$ having the following properties for all states u, v, w:

1. *Null self-distance* $d(u,u) = 0$,
2. *Symmetry* $d(u,v) = d(v,u)$,
3. *Triangle inequality* $d(u,w) \leq d(u,v) + d(v,w)$.

A metric is a generalization of an equivalence relation and includes it as a special case, where

$$d(u,v) = 0$$

defines equivalence of states u and v, $u \equiv v$. This is indeed an equivalence relation, since the properties of (1) reflexivity, (2) symmetry, and (3) transitivity of an equivalence relation follow, respectively, from the properties of (1) null self-distance, (2) symmetry, and (3) triangle inequality of a metric. Thus for a state space equipped with metric the symmetry group is defined as before.

For a state space equipped with metric $d(\quad,\quad)$ some of the transformations might map all states to "nearby" states only. That is, for a given positive number ϵ there might be transformations T such that

$$d(u,T(u)) \leq \epsilon$$

for all states u. Such a transformation is called an *approximate symmetry transformation* of the system. We see that whether a given transformation is an approximate symmetry transformation depends on the value taken for ϵ. The larger the value of ϵ, the more transformations are approximate symmetry transformations in general. The set of all invertible approximate symmetry transformations of a given system for any of its state spaces equipped with metric and for a given ϵ does not in general form a group (there might be trouble with closure), although it does, of course, include the corresponding symmetry group and might include other groups that in turn include the symmetry group.

Since the general theory of approximate symmetry is not very well developed, I do not think it worthwhile to go into many details. It suffices to state that it is possible to define *approximate symmetry groups* for state spaces equipped with metrics, and it is possible to define a measure of *goodness of approximation* for each approximate symmetry group. A system will in general have different approximate symmetry groups with different measures of goodness of approximation. But for the same state space and metric any approximate symmetry group will include as subgroups all approximate symmetry groups of better approximation, and, especially, every approximate symmetry group includes the corresponding symmetry group as a subgroup. Thus the approximate symmetry groups and exact symmetry group of a state space equipped with metric can be arranged in an inclusion chain, with the exact symmetry group at the included end and worsening approximation in the direction of more inclusive groups. An approximate symmetry group does not in general consist of all the invertible approximate symmetry transformations for some value of ϵ. And, perhaps somewhat strangely, for any given value of ϵ an approximate symmetry group might contain transformations that are not approximate symmetry transformations at all.

Again without going into detail, for a state space equipped with metric and an approximate symmetry group of it an *exact symmetry limit* is a metric for which that approximate symmetry group is the exact symmetry group. Since it is a physical system that is being considered, so that the metric has physical significance, an exact symmetry limit of it might correspond to a physically realizable system, to a conceivable but physically unrealizable system, or to nothing conceivable as a physical system at all. Returning to the nuclear physics example of maximalistic use of the symmetry principle presented in Section 4.5, we have charge symmetry as an approximate symmetry. An exact symmetry limit would have charge symmetry as an exact symmetry, i.e., the nuclear force would be acting, while the electromagnetic interaction and the Pauli exclusion principle would be "switched off." Such a situation is conceivable, since by describing it we conceive of it, but it certainly is not physically realizable. In terms of metric, any realistic metric for the actual situation must give nonzero "distances" for pairs of states differing only in that one or more protons in one are replaced by neutrons in the other and vice versa, since the electromagnetic interaction and the Pauli principle distinguish between such states. An exact symmetry limit for charge symmetry gives null "distance" for such pairs, expressing their equivalence in the absence of the electromagnetic interaction and the Pauli principle.

Another term for approximate symmetry is *broken symmetry*. The *symmetry-breaking factor* is whatever factor the (possibly hypothetical) switching off of brings about an exact symmetry limit. In our example the electromagnetic interaction and the Pauli principle constitute the symmetry-breaking factor. Another example is a crystal, which possesses

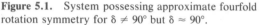

Figure 5.1. System possessing approximate fourfold rotation symmetry for $\delta \neq 90°$ but $\delta \approx 90°$.

broken displacement symmetry. The exact symmetry limit is an infinite crystal, obviously unobtainable in practice. It is the finiteness of the real crystal that breaks the displacement symmetry. Or consider the four-equal-straight-armed figure of Fig. 5.1, where each pair of opposite arms forms a straight line segment and $\delta \neq 90°$ but $\delta \approx 90°$, δ being the angle between the line segments, as shown. This system possesses approximate fourfold rotation symmetry. The symmetry-breaking factor is the difference between δ and $90°$. The exact symmetry limit is the system with $\delta = 90°$ and is realizable.

5.2. SPONTANEOUS SYMMETRY BREAKING

In discussions of symmetry and the symmetry principle, that the effect is at least as symmetric as the cause, the question of "spontaneous symmetry breaking" inevitably arises. There appear to be cases of physical systems where the effect simply has less symmetry than the cause, where the symmetry of the cause is possessed by the effect only as a badly broken symmetry, so that the exact symmetry group of the effect is a subgroup of the symmetry group of the cause, rather than vice versa. In other words, we seem to have broken symmetry with no symmetry-breaking factor. Among those cases are Garrett Birkhoff's (American mathematician, 1911–   ) hydrodynamic "symmetry paradoxes" [A4], and additional examples are noted later in this section. Yet, as we have convinced ourselves, the symmetry principle cannot but hold. The trouble is that what is taken to be the cause is not complete, and the complete cause is less symmetric. What is assumed to be the exact symmetry of the cause is really only an approximate symmetry, although to a very good approximation. The goodness of approximation is so good that we are deceived into believing that it is perfect and overlook the small symmetry-breaking factor that makes the symmetry only approximate. What is in fact usually overlooked is the influence of small, random fluctuations in physical systems. Thus the symmetry principle is saved. The exact symmetry of the cause remains, as it must, the minimal exact symmetry of the effect. But a good approximate symmetry of the cause can be possessed by the effect as a badly broken symmetry.

Just how do small, symmetry-breaking perturbations of a cause affect the symmetry of the effect? What can be said about the symmetry of an effect relative to the *approximate* symmetry of its cause? That depends on the actual nature of the physical system, on whatever it is that links cause and effect in each case. But we can consider the possibilities.

1. *Stability*. The deviation from the exact symmetry limit of the cause, introduced by the perturbation, is "damped out," so that the approximate symmetry group of the cause is the minimal symmetry group of the effect.

2. *Lability*. The approximate symmetry group of the cause is the minimal approximate symmetry group of the effect, of more or less the same goodness of approximation.

3. *Instability*. The deviation from the exact symmetry limit of the cause, introduced by the perturbation, is "amplified," and the minimal symmetry of the effect is only the exact symmetry of the cause (including perturbation), with the approximate symmetry of the cause appearing in the effect as a badly broken symmetry. This is what is commonly called *spontaneous symmetry breaking*. To quote Birkhoff [A4],

> *Although symmetric causes must produce symmetric*
> *effects, nearly symmetric causes need not produce*
> *nearly symmetric effects:* a symmetry problem need
> have no *stable* symmetric solutions.

An example of a stable situation is the solar system. All the planetary orbits lie pretty much in the same plane, and this seems to have been the case for as long as observations have been recorded, or even for as long as the planets have existed, according to the modern theory of the origin of the solar system. This is a symmetry, reflection symmetry with respect to the plane of the solar system. Now if the state of the solar system at any instant is taken as the cause, its state at any later instant is an effect. Thus the solar system exhibits stability for reflection symmetry. In spite of the various and numerous internal (and external) perturbations that the solar system has suffered throughout its history, which might be expected to have broken the reflection symmetry more and more as time went on, the symmetry has obviously been preserved. (Symmetry in processes will be discussed in detail in Chapter 6.)

As an example of lability, consider the DC circuit presented in Section 4.4 as an example of minimalistic application of the symmetry principle (Fig. 4.12). That system possesses twofold rotation symmetry (actually, as explained there, functional permutation symmetry). But it is the ideal system that is symmetric. No real circuit built according to that diagram will be exactly symmetric because of practical limitations on the accuracy of resistors, emf sources, and so on. Yet the effect, the currents and potential drops, always shows twofold rotation symmetry to reasonable accuracy. So the situation is certainly not one of instability. Is it stability or lability? To check this we replace a fixed resistor by a variable one or a fixed emf source by a variable one and gradually break the symmetry of the cause, following the symmetry of the effect with ammeters and voltmeters as we do so. And we find that the symmetry of the effect gets broken as gradually as we break the symmetry of the cause. That proves that the system is labile for twofold rotation symmetry.

For an example of instability we return to the solar system, this time to its origin and evolution. Modern theory has the solar system originating as a rotating cloud of approximate axial symmetry and reflection symmetry with respect to a plane perpendicular to its axis. If that state of what is now the solar system is taken as the cause, the present state can be taken as the effect. And any axial symmetry the proto solar system once had has clearly practically disappeared during the course of evolution, leaving the solar system as we now observe it. The random, symmetry-breaking fluctuations in the original gas cloud grew in importance as the system evolved, until the original axial symmetry became hopelessly broken. Additional examples of instability are Birkhoff's "symmetry paradoxes," referred to previously.

An example exhibiting both stability and instability, although under different conditions, is a volume of liquid at rest in a container. Such a liquid is isotropic; its physical properties are independent of direction. This is a symmetry. The system is stable for isotropy, and our attempts to break the symmetry will be overcome by the liquid, if only it can. A tap on the side of the container introduces anisotropy, which is very soon damped out by the system, and isotropy is regained. A small crystal of the frozen liquid thrown into the liquid also breaks the symmetry, but it soon melts and isotropy returns. However, when the liquid is cooled to below its freezing point, the situation alters drastically. Let us imagine that it is cooled very slowly and evenly and that no isotropy-breaking perturbations are allowed; in other words, assume that the liquid is supercooled. It is then still in the liquid state and isotropic. If we now tap the container or throw in a crystal or otherwise introduce anisotropy, the supercooled liquid will immediately crystallize and thus become highly anisotropic. So in the subfreezing temperature range the system is unstable for isotropy; any anisotropic perturbation is immediately amplified until the whole volume becomes anisotropic and stays that way. If, in spite of our precautions, the liquid should spontaneously crystallize during or after cooling, the reason is that its own internal random fluctuations are sufficient to set off the instability. The cooler the liquid (below its freezing point), the greater its instability. The freezing point is the boundary between the temperature range of stability and that of instability. In fact, it can be defined this way.

Even in situations of lability and instability, if the perturbation is truly a random one, there is a sense in which the symmetry of the unperturbed cause, the exact symmetry limit, nevertheless persists as the minimal symmetry group of the effect. This pivots on the nature of random perturbations, or, more accurately, on their definition. A random perturbation of an unperturbed cause is defined, for our purpose, as a fluctuating perturbation such that, as the number of observations increases, the symmetry of the *totality* of observed states of the perturbed cause tends to that of the unperturbed cause. Thus, taking the totality of states of a cause as "the cause" in the symmetry principle, we have that for a random perturbation the totality of states of the effect, as "the effect" of the symmetry principle, tends to exhibit at least the symmetry of the unperturbed cause as the number of observations increases.

As an illustration consider the system consisting of a current of air blowing against the edge of a wedge, with the direction of the air current far before the wedge parallel to the bisector of the wedge, as in Fig. 5.2. Take as the cause the wedge and the incident air current a large distance before the wedge, and the air flow around the wedge as the effect. The cause possesses reflection symmetry with respect to the plane bisecting the angle of the wedge. So we expect the air flow above the wedge to be the mirror image of the air flow under the wedge. And, in fact, that is what happens, as long as the velocity of the incident air current is sufficiently low so that the flow is laminar. However, at sufficiently high velocities vortices form at the edge and are shed downstream. They are not produced symmetrically, but rather alternately at one side and the other. (This periodic phenomenon is the mechanism of production of

Figure 5.2. Air current blowing against edge of wedge, viewed in cross section.

"edge tones," the acoustic source of woodwind instruments.) The effect then apparently has lower symmetry than the cause, since random fluctuations in the direction of incident air flow make the reflection symmetry of the cause only an approximate symmetry. At low incident-flow velocities the situation is labile, and the resulting laminar flow around the wedge is approximately symmetric. (Underblowing a woodwind instrument produces no tone.) At high velocities the situation is unstable, and the vortical flow around the wedge lacks reflection symmetry altogether. But if the vortical flow is photographed sufficiently often at random times, the total collection of photographs will exhibit reflection symmetry; for every flow configuration photographed a mirror-image configuration will also appear among the pictures.

I would like to add that it seems to me that the recently developed "theory of catastrophes" [M48, M35, M43] is most likely the suitable mathematical language for the description of instabilities of the kind we have been discussing. See also [M21].

The situation, discussed in Section 4.6, where the vacuum state of a quantum theory is less symmetric than the rest of the theory, can be viewed as broken symmetry, with the symmetry of the theory without the vacuum state taken as the exact symmetry limit and the vacuum state as the symmetry-breaking factor. However, such a situation is sometimes referred to as *spontaneous* symmetry breaking. This nomenclature is misleading, and the attribution of spontaneity is not justified. The vacuum state is an obvious component of the cause for such theories, and its effect in reducing the symmetry of the total cause from that of the exact symmetry limit is in no sense small, nor is it (or should it be) liable to be overlooked. What might justifiably be called *spontaneous* symmetry breaking in this connection is a higher-level theory, a super theory, which is supposed to explain the first theory, including the lower symmetry of the vacuum state compared with that of the rest of the theory, or at least it is supposed to explain the vacuum state. If the super explanation involves instability and amplification of perturbations, so that the vacuum state, as an effect of the super theory, comes out possessing a lower degree of symmetry than that of the unperturbed super cause, whatever that is, then the use of the term "spontaneous" is reasonable and consistent with our previous discussion. A situation like that is described in Section 6.5.

CHAPTER SIX

Symmetry in Processes, Conservation, and Cosmic Speculations

In this chapter we discuss the concepts of the laws of nature and their symmetry, including time-reversal symmetry. We then explain the analysis of nature into laws of nature and initial conditions and discuss symmetry of initial and final conditions, leading to the equivalence and symmetry principles for natural processes in isolated physical systems and to the general and special symmetry evolution principles for such systems. The latter two principles are both concerned with the nondecrease of degree of symmetry during evolution. We obtain an explanation for the empirical observation that macrostates of stable equilibrium of physical systems are often especially symmetric. Symmetry and entropy are related to each other. Then we discuss conservation and its relation to symmetry of the laws of nature. And finally we indulge in cosmic speculation about it all.

The previous chapters are prerequisite. The prerequisite scientific background for Chapters 4 and 5, with the addition of some kinetic theory and some familiarity with the concepts of macrostate and microstate, should suffice for this chapter. Acquaintance with the "big bang" and some elementary particle physics should also prove useful. The prerequisite mathematics for Section 6.4 is calculus, including partial differentiation.

6.1. SYMMETRY OF THE LAWS OF NATURE

Laws of nature are what in Chapter 4 we called "causal relations in systems" when the "systems" are processes in physical systems, with the "cause subsystem" being the initial state of the physical system and the "effect subsystem" its final state. (Note and beware of the two ways the term "system" is being used here!) Yes, it is the dynamic process, the temporal evolution of the physical system, that is taken as the "system," while the state of the physical system at any initial time is the "cause subsystem" and the state at any final time is the "effect subsystem." The *laws of nature,* or *laws of evolution,* are the causal relation between cause and effect subsystems in such cases, the causal relation between initial and final states of physical processes. Equivalently, and in more familiar language, the laws of nature can be viewed as the natural temporal development of physical systems from initial states to final states. See [M13].

It must be emphasized that we are considering only isolated physical systems, where by "isolated" we mean that there is no interaction with the rest of the world, that the physical systems evolve under internal influences only. This is implied by the existence of causal relation, cause subsystem, and effect subsystem, with initial states leading to unique final states.

We have all heard of various symmetries and approximate symmetries that are ascribed to the laws of nature, such as special relativistic symmetry, $SU(3)$ symmetry, isospin symmetry, charge symmetry, particle–antiparticle-conjugation (C) symmetry, time-reversal (T) symmetry, space-inversion (P) symmetry, CP symmetry, CPT symmetry. (We cannot go into a description of them here.) There are two points of view about the meaning of symmetry of the laws of nature.

1. A scientist (with a laboratory) investigates nature and discovers laws. Another scientist (with his or her laboratory), related to the first by some transformation, also investigates nature and discovers laws. If these laws are the same as those discovered by the first scientist (or if a certain subset of these is the same as a certain subset of those), and if, moreover, this is true for all pairs of scientists related by the transformation, then the transformation is a symmetry transformation of the laws of nature (or of a certain subset of them). This is called the passive point of view.

2. If among all processes that can be conceived as occurring naturally (or a certain subset of them) any pair related by a certain transformation are either both allowed or both forbidden by nature (or, in quantum

phenomena, both have the same transition amplitude), then that transformation is a symmetry transformation of the laws of nature (or of a certain subset of them). This is called the active point of view.

For example, the symmetry of the special theory of relativity can be expressed passively: (1) All pairs of scientists with arbitrary spatiotemporal separation, arbitrary relative orientation, and arbitrary constant rectilinear relative velocity discover the same laws of nature, including the speed of light. Or it can be put actively: (2) Among all processes that can be conceived as occurring naturally, including the propagation of a light signal over a certain distance during a certain time lapse, any two that are identical except for arbitrary spatiotemporal separation, arbitrary relative orientation, and arbitrary constant rectilinear relative velocity are either both allowed or both forbidden by nature (or have the same transition amplitude).

Another example, CP symmetry, which seems to be valid in all of nature except for a certain class of weak or superweak interactions among elementary particles, can be formulated passively: (1) All pairs of scientists that are particle–antiparticle conjugates and space-inversion images of each other discover the same laws of nature (with certain exceptions). The transformation involved here is converting a scientist into an antiscientist, i.e., replacing all the protons, neutrons, electrons, and other particles comprising the scientist and his or her laboratory with antiprotons, antineutrons, positrons, and other antiparticles, respectively, and then point inverting the antiscientist (or first invert and then conjugate; the results are the same). Since antiscientists are not at present physically realizable (at least not in the sense just defined), the passive point of view seems absurd, so one might prefer the active one: (2) Among all processes that can be conceived as occurring naturally (with certain exceptions), any two that are identical except for particle–antiparticle conjugation and space inversion are either both allowed or both forbidden by nature (or have the same transition amplitude).

The two points of view are called "passive" and "active," respectively, because of the nature of the transformations under which the laws of nature are symmetric. A transformation is called "passive" if it does not change the world but changes the way the world is observed, i.e., it changes the frame of reference. For a geometric transformation, for example, this would be a change of coordinate system. A transformation is called "active" if it does not change the observer but changes the rest of the world. The same reference frame is used for a changed world. In the geometric case the locations of events would be changed with respect to a fixed coordinate system.

I prefer the active (second) point of view for considering symmetry of the laws of nature. The main reason is that, as we saw in the example concerning *CP* symmetry, transformed reference frames (observers) are not always physically realizable. From the active point of view a symmetry transformation of the laws of nature can be concisely defined as a transformation for which the transformed result of an experiment is the same as the result of the transformed experiment for all experiments (or for some subset of experiments, in which case we have a symmetry transformation of a subset of the laws of nature).

Here is a precise, diagrammatic formulation of symmetry of the laws of nature. The laws of nature, or laws of evolution, are expressed by a transformation, denoted by *N* (for "nature"), of the state space of every physical system into itself, a mapping giving as the image of any state the final state evolving naturally from the object state as initial state and thus exhibiting the causal relation between initial and final states. See Fig. 6.1. (Since not all states are necessarily obtainable as final states, there may be states that do not serve as images in this mapping. How then, you may well ask, are such states obtained to serve as initial states? They are set up by tampering with the system, i.e., they are the final states of processes during which the system is *not isolated*!) So if *u* denotes the initial state of any physical system, $N(u)$ is the final state that is the result of the evolution of the system from initial state *u*. We can also use the notation

$$u \xrightarrow{N} N(u)$$

to make the transformation graphically look more like a process. This

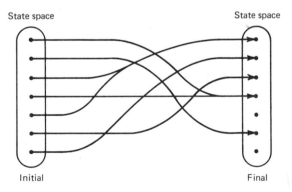

State space   State space

Initial   Final

Figure 6.1. Laws of evolution for isolated physical system are mapping of system's state space into itself, where image of any state is final state evolving from object state as initial state.

discrete, initial → final model of evolution will best serve the purposes of our discussion. Continuous evolution is obtained in the limit of infinite reiteration of infinitesimal discrete steps.

Let $\mathcal{T}$ denote any transformation of every physical system, such as rotation, spatial displacement, reflection, particle–antiparticle conjugation. The image of state u of any physical system is state $\mathcal{T}(u)$ of the same system, or

$$u \xrightarrow{\mathcal{T}} \mathcal{T}(u),$$

although we are not indicating a process this time.

The process

$$u \xrightarrow{N} N(u)$$

is (or could be) the running of an experiment, where u denotes the initial experimental setup and $N(u)$ denotes the result. Transforming the experimental result $N(u)$ by $\mathcal{T}$ gives image state $\mathcal{T}N(u)$, or

$$N(u) \xrightarrow{\mathcal{T}} \mathcal{T}N(u).$$

Now consider the transformed experiment, the image of the above experiment under $\mathcal{T}$. Its initial setup is obtained by transforming state u to state $\mathcal{T}(u)$, or

$$u \xrightarrow{\mathcal{T}} \mathcal{T}(u).$$

Starting from initial state $\mathcal{T}(u)$, the experiment takes place, proceeding according to the laws of nature, and yields the result $N\mathcal{T}(u)$, or

$$\mathcal{T}(u) \xrightarrow{N} N\mathcal{T}(u).$$

Combining these results, our definition of a symmetry transformation of the laws of nature, a transformation $\mathcal{T}$ for which for all experiments the transformed result is the same as the result of the transformed experiment, now becomes

$$\mathcal{T}N(u) = N\mathcal{T}(u)$$

for all states u, or diagrammatically as in Fig. 6.2. Since this holds for all states u, we have the definition of a symmetry transformation of the laws of nature: The transformation under consideration commutes with

Figure 6.2. Symmetry transformation of laws of nature is any transformation $\mathcal{T}$ for which diagram is valid for all states u. N is evolution transformation.

the evolution transformation,

$$\mathcal{T}N \;=\; N\mathcal{T}$$

(see Section 3.2). (For a symmetry transformation of a subset of the laws of nature things must be formulated accordingly in terms of all states u belonging to a certain subspace of state space.)

The fundamental point underlying these definitions and formulations is this: A symmetry of the laws of nature is an indifference of the laws of nature. For a transformation to be a symmetry transformation of the laws of nature, the laws of nature must ignore some aspect of physical states, and the transformation must affect that aspect only. A pair of initial states related by such a transformation are treated impartially by the laws of nature; they evolve into a pair of final states that are related by precisely the same transformation. The laws of nature are blind to the difference between the two states, which is then preserved during evolution and reemerges as the difference between the two final states.

As an example, consider spatial-displacement symmetry of the laws of nature, meaning that the laws of nature are the same everywhere. If we perform two experiments that are the same except for one being here and the other being there, they will yield outcomes that are the same except for one being here and the other there, respectively. And this is found to be valid for all experiments and for all heres and theres. For a picture see Fig. 6.3.

These insensitivities of the laws of nature remind us of the "impotencies" of scientific theories, discussed in Section 4.2. Indeed, since theories are expressions of the laws of nature, all symmetries of the laws of nature must appear as impotencies of theories, if the theories are to express the laws of nature faithfully. Any subset of the laws of nature will possess the symmetry of the full set and might have additional symmetry. Partial theories, those theories that are concerned with only part of the laws of nature, might accordingly have additional impotencies.

For example, since the laws of nature are symmetric under spatial displacements, no scientific theory may accept absolute position as input. For examples of partial theories, the gas laws are concerned with macroscopic states of gases, and Kirchhoff's laws are concerned with electric currents. In the former case, although the laws of nature do have something to say about which gas molecules go where and when, the gas laws ignore microscopic aspects of the gas. In the latter case, although the laws of nature do apply to the motion of each of the moving electrons comprising the electric current, Kirchhoff's laws ignore such details.

The set of invertible symmetry transformations of the laws of nature, i.e., the set of all invertible transformations commuting with the evolution transformation, is easily shown to form a group. We call this group *the symmetry group of the laws of nature*.

Another possible symmetry of the laws of nature, very different in kind from the others, is temporal-inversion, or time-reversal, symmetry, also called reversibility. The time-reversal transformation acts on a process by replacing each state by its time-reversal image and reversing the temporal order of events. A moving picture projected in reverse is a good model of time reversal. The time-reversal image of a state depends on the type of state involved; for classical mechanics it is obtained merely by reversing the senses of all velocity vectors. The question then is whether the image process, leading from the time-reversed final state to the time-reversed initial state, is the process that would evolve naturally from the time-reversed final state. If this is so for all processes, the laws of evolution possess time-reversal symmetry, or are reversible. This is illustrated in Fig. 6.4.

In symbols the object process is

$$u \xrightarrow{\ N\ } N(u).$$

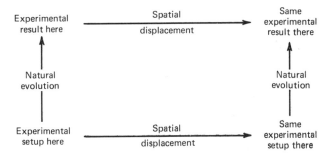

Figure 6.3. Spatial-displacement symmetry of laws of nature. Diagram is valid for all experiments and for all heres and theres.

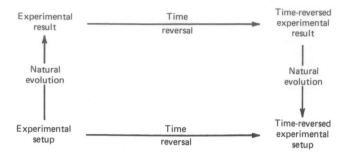

Figure 6.4. Time-reversal symmetry, or reversibility, of laws of nature, if diagram is valid for all experiments.

The time-reversal image of the final state $N(u)$ is $TN(u)$,

$$N(u) \xrightarrow{T} TN(u),$$

where T stands for time reversal. The natural evolution from state $TN(u)$ is

$$TN(u) \xrightarrow{N} NTN(u).$$

And if the final state of this process is the time-reversal image of the initial state of the object process, i.e., if

$$NTN(u) = T(u)$$

for all states u, we have time-reversal symmetry, the definition of which is then

$$NTN = T.$$

Note how this differs from the definition, derived previously, of a symmetry transformation of the usual kind.

With a little thought it will be clear that for the laws of nature to possess time-reversal symmetry, i.e., for the definition we just derived to be valid, it is necessary (although not sufficient) that the evolution transformation be both one-to-one and onto, that N be invertible. So, if we like, we can put the definition we just obtained in the form

$$TN = N^{-1}T.$$

At present time-reversal symmetry, or reversibility, seems to be valid in nature for the laws of evolution of almost all microscopic systems, more precisely, for all microscopic systems that do not contain neutral kaons. On the other hand, it is invalid for almost all macroscopic systems.

Problems

1. No real physical system is isolated. Discuss possible kinds of external influence. Consider the question of stability against external influence. How is it that we discover laws of nature, although our experimental systems are not precisely isolated? Speculate about what might happen in perfectly isolated physical systems.

2. Prove that the set of invertible symmetry transformations of the laws of nature forms a group.

3. What underlies the difference between time-reversal symmetry and symmetries whose symmetry transformations commute with the evolution transformation?

6.2. SYMMETRY OF INITIAL AND FINAL CONDITIONS, THE GENERAL SYMMETRY EVOLUTION PRINCIPLE

. In the way we usually analyze nature, the laws of nature do not exhaust all of what is going on. The laws of nature do indeed determine how any process, once started, will evolve and what its outcome will be. But it is the initial conditions, the situation at any given single instant, that determine just which process is to take place. And, with an important exception discussed in Section 6.5, the laws of nature do not determine initial conditions. Or perhaps we might say that initial conditions are whatever nature allows us to have control over, at least in principle, or, perhaps even better, whatever nature prefers not to be bothered with. Thus science usually makes a sharp distinction between the laws of nature, on the one hand, and initial conditions, on the other.

For an example in classical mechanics, consider a set of bodies interacting only with each other. The laws of evolution are Newton's laws of motion (Isaac Newton, English philosopher and mathematician, 1642–1727) and the interaction among the bodies. We may arbitrarily specify the positions, momenta, orientations, and angular momenta of all the bodies at a given time but no more than that. So these are initial conditions.

Since the laws of nature are beyond our control, the initial conditions—the initial state of the physical system—uniquely determine for us the process and its outcome—the final state of the physical system. Thus, as described in the preceding section, initial conditions and outcome are in causal relation. Taking the whole process as our "system," the set of initial conditions and the set of final conditions are cause and effect subsystems, respectively.

(I again emphasize that we are considering only isolated physical systems, in which initial states do indeed uniquely determine the outcomes of processes, in which there are no outside influences messing things up.)

Symmetry of the laws of nature (time-reversal symmetry excluded) determines an equivalence relation in the state space of a system: A pair of states is equivalent if and only if they are indistinguishable by the laws of nature, i.e., if and only if some element of the symmetry group of the laws of nature carries one into the other. Thus the symmetry group of the laws of nature determines a decomposition of state space into equivalence subspaces, the physical significance of which is that the members of each equivalence subspace are all those states and only those that are indistinguishable by the laws of nature. Then, following reasoning similar to that of Section 4.2, we obtain *the equivalence principle* for processes in isolated physical systems:

> *Equivalent states, as sets of initial conditions,*
> must *evolve into equivalent states, as sets of*
> *final conditions, while inequivalent states* may
> *evolve into equivalent states.*

See Fig. 6.5.

We can therefore count on and exploit the equivalence principle, whether we are thinking in terms of the laws of nature, part of the laws of nature, a scientific theory, or a partial scientific theory.

For an example of the equivalence principle in the present context, since the laws of nature possess spatial-displacement symmetry, initial states that differ only in position evolve into final states that differ only in position.

Or all known laws of nature are symmetric under interchange of elementary particles of the same species, so such particles are, as far as we know, inherently indistinguishable. Two initial states differing only in the interchange of, say, two electrons cannot, by any known process, lead to distinguishable final states.

Or certain laws of nature might apply to macroscopic states of a system and be indifferent to microstates, as long as the latter correspond to the same macrostate. Then two different initial microstates corresponding to the same initial macrostate will evolve into final microstates that cor-

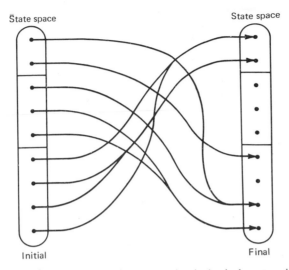

Figure 6.5. Decomposition of state space of isolated physical system into equivalence subspaces determined by symmetry group of laws of nature. According to the equivalance principle equivalent states always evolve into equivalent states, while inequivalent states may also do so.

respond to the same final macrostate. Among these are the gas laws. Kirchhoff's laws care nothing about the composition and structure of resistors and emf sources, as long as they have the resistance and emf they are supposed to have. Newton's laws are obviously indifferent to many aspects of the states to which they apply (color, odor, etc.).

Let us call the equivalence relation defined by the symmetry group of the laws of nature "initial equivalence" for convenience, and we now define another, possibly different equivalence relation, which we call "final equivalence," as follows. A pair of states is final-equivalent if and only if the pair of states evolving from them is initial-equivalent, i.e., is equivalent with respect to the laws of nature. Initial equivalence implies final equivalence, by the equivalence principle, since initial-equivalent states must evolve into initial-equivalent states and are thus final-equivalent. However, states that are not initial-equivalent may also evolve into initial-equivalent states and thus be final-equivalent. So in the decomposition of the state space of a physical system into final-equivalence subspaces, each final-equivalence subspace consists of one or more initial-equivalence subspaces in their entirety. See Fig. 6.6.

When considering symmetry in processes, the process is the "system" in the general sense of the symmetry and equivalence principles, and the set of all possible processes of a physical system is the "state space" in

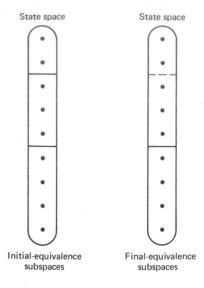

State space

State space

Initial-equivalence
subspaces

Final-equivalence
subspaces

Figure 6.6. Decomposition of state space of isolated physical system into initial-equivalence subspaces and into final-equivalence subspaces for example of Fig. 6.5. Each final-equivalence subspace consists of one or more initial-equivalence subspaces.

the same general sense. Transformations of this "state space" act on processes. However, as we have seen, the set of all possible processes stands in one-to-one correspondence with the set of states of the physical system, since every state, as initial state, initiates a unique process, and every process starts with some initial state. So, to keep things as uncomplicated as possible (relatively speaking, of course), we apply our transformations to the physical system's state space rather than to process space. As we transform the states about, we keep in mind that each one drags around with it the process it initiates, just like mice running around in a cage, each with its tail.

Then since, as we saw above, initial states are the cause subsystem and final states the effect subsystem for processes in isolated physical systems, we have that the symmetry group of the cause (see Section 4.3) is the group of all invertible transformations of state space that preserve initial-equivalence subspaces. And the symmetry group of the effect is the group of all invertible transformations preserving final-equivalence subspaces. And since, as we saw in the paragraph before last, any initial-equivalence-preserving transformation is also final-equivalence-preserving:

The "initial" symmetry group (that of the cause)
is a subgroup of the "final" symmetry group
(that of the effect).

This is *the symmetry principle* for processes in isolated physical systems. In this sense we can say:

> *For an isolated physical system the degree of symmetry cannot decrease as the system evolves, but either remains constant or increases.*

(For degree of symmetry see Section 3.6.) We call this result *the general symmetry evolution principle.* The adjective "general" is included in the name of the principle because the principle is derived from very fundamental considerations with no additional assumptions. Thus it is indeed general, so general, in fact, as to make it rather useless, as we will explain and remedy in the next section. Our motive for deriving it is theoretical, not utilitarian; we should see where our fundamental considerations lead us when applied to the evolution of isolated physical systems.

Problem

We stated that states differing only in the interchange of two elementary particles of the same species are indistinguishable by the laws of nature. And, indeed, we can think of no way of distinguishing among such states. However, due to the spatial-displacement symmetry of the laws of nature, states differing only in their location are also indistinguishable by the laws of nature. Is there a way of distinguishing among such states? If so, consider the lack of analogy with the former case. Are we not running into difficulty? Try to resolve the difficulty.

6.3. THE SPECIAL SYMMETRY EVOLUTION PRINCIPLE AND ENTROPY

The reason the general symmetry evolution principle, although perfectly valid, is not very useful is that, when we consider the evolution of an isolated physical system, we consider the sequence of states it passes through and their symmetry and are usually not interested in the entire state space of the system, in terms of whose transformations the "initial" and "final" symmetry groups were defined in the preceding section. The symmetry group of a single state is the group of all invertible transformations of the physical system that carry the state into equivalent states, where equivalence is with respect to the laws of evolution. Thus the symmetry group of a state is just the group of permutations of all members of the equivalence subspace in state space to which the state belongs (where we generalize suitably for equivalence subspaces with discretely or continuously infinite populations, or, to use the terms presented in Section 1.1, of denumerably or nondenumerably infinite orders). So the degree of symmetry of a state can be measured by the population of its equivalence subspace: The more states equivalent to it, the higher its degree of symmetry.

What can be stated about the degrees of symmetry of the sequence of states through which an isolated system evolves? Or equivalently, what do we know about the populations of the sequence of equivalence subspaces to which those states belong? In spite of the general symmetry evolution principle we know nothing about this in general. We can, however, obtain a result by making the assumption of nonconvergent evolution: Different states always evolve into different states. Then the population of the equivalence subspace of a final state is at least equal to that of the initial state that evolved into the final state, since, by the equivalence principle, all members of the initial state's equivalence subspace evolve into members of the final state's equivalence subspace and, by the nonconvergence assumption, the number of the latter equals the number of the former. Moreover, additional states, inequivalent to the initial state, may also evolve into members of the final state's equivalence subspace (convergence of equivalence subspaces), and these members will be distinct from those we just counted. Thus:

As the system evolves, the populations of the
equivalence subspaces of the sequence of states
through which it passes cannot decrease but
either remain constant or increase.

Or, equivalently,

The degree of symmetry of the state of an
isolated system cannot decrease during evolution
but either remains constant or increases.

See [M36]. We call this result *the special symmetry evolution principle*. See Fig. 6.7. (Compare the special and the general symmetry evolution principles; both derive from the equivalence principle for processes, while the special principle involves the additional assumption of nonconvergent evolution.)

The symmetry group of the final state, the group of permutations of

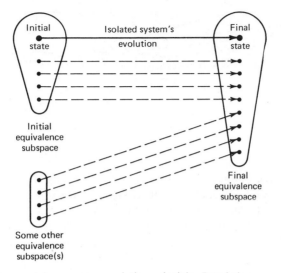

Figure 6.7. The special symmetry evolution principle. Population, measuring degree of symmetry, of final equivalence subspace at least equals that of initial equivalence subspace. This follows from the equivalence principle and assumption of nonconvergent evolution. Solid arrow indicates actual process being considered, while dashed arrows indicate other possible processes.

all members of its equivalence subspace, clearly includes as a subgroup the symmetry group of the initial state, the permutation group of *its* equivalence subspace (all groups being considered abstractly). Actually it includes as subgroups the symmetry groups of all states evolving into states equivalent to the final state. Furthermore, it even includes as a subgroup the direct product of the permutation groups of all distinct equivalence subspaces of states evolving into states equivalent to the final state (or, we could say, of all distinct equivalence subspaces converging to the final equivalence subspace).

The assumption of nonconvergent evolution (of states, of course, not of equivalence subspaces) seems to be valid in nature for microscopic processes, at least as far as we now know. However, it is not valid for macroscopic processes, which is just fine, since the essence of the macroscopic character of a macrostate is that a macrostate is an equivalence subspace of microstates, and such equivalence subspaces are just what we are dealing with here.

Evolution with constant degree of symmetry is typical of microscopically considered systems, systems about which sufficient information is available so that they need not be treated by statistical methods. The mechanism of evolution with constant degree of symmetry is, of course,

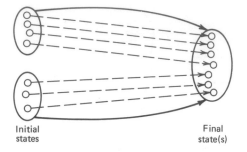

Figure 6.8. Convergent evolution of macrostates (equivalence subspaces of microstates). Circles denote microstates; ellipses denote macrostates. Dashed arrows represent (nonconvergent) evolution of microstates, while solid arrows represent convergent evolution of macrostates.

Initial states

Final state(s)

nonconvergence of equivalence subspaces. Reversibility, or time-reversal symmetry, of the laws of evolution implies evolution with constant degree of symmetry. (If the degree of symmetry increased, the time-reversed process would take place with decreasing degree of symmetry, which is forbidden by the special symmetry evolution principle.) Thus evolution with constant degree of symmetry is a necessary (but not sufficient!) condition for reversibility.

As an example of evolution with constant degree of symmetry, consider a system consisting of a few nucleons and pions about which sufficient information is available. The symmetry transformations of the laws of evolution, which define state equivalence, are for this example spatial and temporal displacements, rotations, boosts, particle–antiparticle conjugation, spatial (point or plane) reflection, and global isospin phase transformations (which are not described here). The evolution of this system is reversible, since its laws of evolution are time-reversal symmetric (quantum time-reversal symmetry). Thus the system evolves with constant degree of symmetry. As it evolves, the populations of the equivalence subspaces of the sequence of states constituting the process are all equal.

Evolution with increasing degree of symmetry is typical of macroscopically considered systems, systems about which insufficient information is available so that they must be treated by statistical methods [M4]. From the special symmetry evolution principle it follows immediately that evolution with increasing degree of symmetry implies irreversibility, or time-reversal asymmetry, of the laws of evolution. The mechanism of evolution with increasing degree of symmetry is convergence of equivalence subspaces, i.e., convergent evolution of macrostates. See Fig. 6.8. So, thinking macroscopically of macrostates, macrostate spaces, and macroscopic evolution transformations N, such N's are not invertible.

An example of evolution with increasing degree of symmetry is a confined, isolated gas considered macroscopically. Let the gas initially occupy any part of the container and have volume V_i and temperature θ_i.

The final macrostate is a homogeneous distribution of gas in the whole container with volume V_f and temperature θ_f. All initial macrostates with the same V_i and θ_i (but with different parts of the container occupied) lead to the same final macrostate (and there are also initial macrostates with different V_i and θ_i that lead to the same final macrostate as well). The macroscopic laws of evolution are blind to the differences among microstates corresponding to the same macrostate. Thus each macrostate is an equivalence subspace of microstate space. In microscopic terms there is convergence of equivalence subspaces during evolution; not only do equivalent initial microstates (corresponding to the same V_i and θ_i and the same occupied part of the container) lead to equivalent final microstates, but there are also inequivalent initial microstates (corresponding to the same V_i and θ_i but different occupied parts of the container, or to different V_i and θ_i) that lead to equivalent final microstates. Since microstates do not converge during evolution, the population of the final equivalence subspace, which is the number of microstates corresponding to the final macrostate, is larger (actually very much larger) than the population of the initial equivalence subspace.

The final macrostate in this example is a state of stable equilibrium of the system. As such it is the outcome of evolution from a very large number of initial macrostates. By the special symmetry evolution principle its symmetry group (the group of permutations of all microstates corresponding to it) must include as a subgroup the direct product of the symmetry groups of all those initial macrostates, as we observed above, and must therefore be of relatively high order. In other words:

> *The degree of symmetry of a macrostate of*
> *stable equilibrium must be relatively high.*

This is a general theorem, independent of the example in the context of which it was derived. Although this symmetry is with respect to permutations of equivalent microstates, it can have macroscopic manifestations. In the example the final state is homogeneous, i.e., symmetric under all permutations of subvolumes of the gas. The theorem is in accord with observation [M45].

But what about those systems that insist on evolving toward reduced symmetry in spite of our theoretical arguments? For example, the solar system is thought to have evolved from a state of axial and reflection symmetry to the present, less symmetric state of only reflection symmetry. And the evolutionary development of plants and animals seems to be toward a reduction of symmetry. The answer is that those systems do not fulfill the conditions of our argument. They are either only approximately symmetric (while our argument was based on exact symmetry) or not isolated (i.e., subject to external influences, which we excluded) or both and are not stable against such perturbations. Since no symmetry of a macroscopic physical system is exact and no physical system is absolutely isolated (both conditions being convenient and useful idealizations), the crucial point is that of stability; a sufficiently stable system will behave as we described, will evolve with nondecreasing symmetry, and, if macroscopic, will have relatively highly symmetric states of stable equilibrium.

Under the correspondence

$$\text{degree of symmetry} \leftrightarrow \text{entropy}$$

the special symmetry evolution principle and the second law of thermodynamics are isomorphic. Both are concerned with the evolution of isolated systems; both state that a quantity, a function of macrostate, cannot decrease during evolution. So, following good science practice, we assume that a fundamental relation exists between entropy and degree of symmetry and look for a functional relation giving the value of one as a strictly monotonically ascending function of the other. This is already known, however, from the statistical definition of entropy,

$$S = k \log W,$$

where W is the number of microstates (the population of the equivalence subspace of microstate space) corresponding to the macrostate for which the value of the entropy, S, is thus defined. (k is the Boltzmann constant. This definition holds for the case of nonconvergence of microstates during evolution.) As mentioned, W can be used to measure the degree of symmetry of the macrostate, and the function $k \log W$ is indeed a strictly monotonically ascending function of W. See [M40].

Thus entropy and degree of symmetry for isolated systems either remain constant together or increase together, with a concomitant constancy or decrease in the degree of order for the system. Note that *disorder* stands in positive correlation with symmetry. Total chaos is maximal symmetry. Organization implies and is implied by reduction of degree of symmetry. The hypothetical "heat death" or "entropy death" of the universe can also be called its "symmetry death."

Problem

The subject of the last paragraph of this section deserves further contemplation, I think. Consider various implications of it. Compare with Pierre

Curie's (French physicist and chemist, 1859–1906) article [A17], where he states that it is asymmetry that produces phenomena.

6.4. CONSERVATION

There is more to be said about symmetry of the laws of nature, and that concerns conservation. As you most likely know, there are a number of conservations (called "conservation laws" in the past) that hold for isolated systems. The best known of them are conservation of energy, conservation of linear momentum, conservation of angular momentum, and conservation of electric charge. What is meant is that, if the initial state of any isolated physical system is characterized by having definite values for one or more of these quantities, then any state that evolves naturally from that initial state will have the same values for these quantities. See [M12].

For example, in a particle scattering experiment the vector sum of the linear momenta of all the particles before the scattering has taken place equals the vector sum of the linear momenta of all the particles after they have interacted with each other. In addition, the total charge of the particles before collision equals their total charge afterwards. And similarly for total angular momentum and total energy. These conservations hold even if particles are produced or annihilated during the process!

It turns out that each conservation is intimately and fundamentally related to a group of symmetry transformations of the laws of nature. Although there is considerable theoretical understanding of this relation, there is still much room for further investigation of it. A theoretical discussion of this relation is beyond the scope of this book. However, let us mention the symmetries of the laws of nature that are related to the conservations referred to above and show (for three of the four) how each conservation can be derived analytically from its related symmetry for a simple mechanical system.

Conservation of energy is related to symmetry of the laws of nature under the group of temporal-displacement transformations, to the fact that the laws of nature do not change with time. This is called temporal homogeneity of the laws of nature. It means that for every naturally allowed process all processes that are identical with that one except for their occurring at different times are also allowed by nature.

Consider the nonrelativistic system of a single point particle of mass m moving in one dimension in a potential V that is a function only of the particle's coordinate x. Since the potential has no explicit dependence on time t, the laws governing this system's evolution are temporal-displacement symmetric. Denote the particle's velocity by $\dot{x}$ and its acceleration by $\ddot{x}$. Its total energy is

$$E = \tfrac{1}{2}m\dot{x}^2 + V.$$

The time rate of change of the total energy is

$$\frac{dE}{dt} = m\dot{x}\ddot{x} + \frac{dV}{dx}\dot{x}.$$

The derivative dV/dx is, by the definition of potential, the negative of the force on the particle, which in turn, by Newton's second law, equals $-m\ddot{x}$, giving

$$\frac{dE}{dt} = m\dot{x}\ddot{x} - m\ddot{x}\dot{x} = 0.$$

Thus the total energy does not change with time; it is conserved.

Conservation of linear momentum is related to symmetry of the laws of nature under the group of spatial-displacement transformations, to the fact that the laws of nature are the same everywhere. This is called spatial homogeneity of the laws of nature. It means that for every naturally allowed process all processes identical to that one but taking place at different locations are also allowed by nature.

Consider a nonrelativistic one-dimensional system consisting of a number of point particles, labeled by i, j, of masses m_i. Let x_i denote the coordinate of the ith particle, $\dot{x}_i$ its velocity, and $\ddot{x}_i$ its acceleration. Let the particles interact via a potential that depends only on the differences of the particles' coordinates, $x_i - x_j$. This potential is independent of the location of the system relative to the coordinate origin, so the laws governing the system's evolution are spatial-displacement symmetric. The total linear momentum of the system is

$$P = \sum_i m_i\dot{x}_i.$$

The time rate of change of the total momentum is

$$\frac{dP}{dt} = \sum_i m_i\ddot{x}_i.$$

The term $m_i\ddot{x}_i$ is the force on the ith particle, which equals $-\partial V/\partial x_i$ by the definition of potential. Thus

$$\frac{dP}{dt} = -\sum_i \frac{\partial V}{\partial x_i}$$

$$= -\sum_i \sum_{j \neq i} \frac{\partial V}{\partial(x_i - x_j)}$$

$$= 0,$$

since each pair of terms with i and j interchanged cancel in the double sum. Thus the total momentum does not change with time; it is conserved.

Conservation of angular momentum is related to symmetry of the laws of nature under the group of all rotations about all axes through a point, to the fact that the laws of nature are the same in all directions. This is called isotropy of the laws of nature. It means that for every naturally allowed process all processes that are identical with that one but have different spatial orientations are also allowed by nature.

Consider the nonrelativistic system of a single point particle of mass m moving in a plane in a central potential, i.e., a potential depending only on the particle's distance from the origin. Since this potential is independent of the particle's orientation relative to the coordinate axes, the laws governing the system's evolution are rotation symmetric about the axis through the origin and perpendicular to the plane of the system. Let (x,y) denote the particle's coordinates. Then its distance from the origin is

$$r = \sqrt{(x^2 + y^2)}.$$

Denote the x-component of the particle's velocity by $\dot{x}$, the y-component of its velocity by $\dot{y}$, the x-component of its acceleration by $\ddot{x}$, and the y-component of its acceleration by $\ddot{y}$. The angular momentum of the particle with respect to the origin is

$$M = m(x\dot{y} - y\dot{x}).$$

The time rate of change of the angular momentum is

$$\frac{dM}{dt} = m(x\ddot{y} - y\ddot{x}).$$

The expression $m\ddot{y}$ is the y-component of the force on the particle and equals $-\partial V/\partial y$. Similarly, $m\ddot{x}$, the x-component of the force, equals

$-\partial V/\partial x$. Now

$$\frac{\partial V}{\partial y} = \frac{dV}{dr}\frac{\partial r}{\partial y} = \frac{y}{r}\frac{dV}{dr},$$

and similarly,

$$\frac{\partial V}{\partial x} = \frac{x}{r}\frac{dV}{dr}.$$

Therefore

$$\frac{dM}{dt} = \left(-\frac{xy}{r} + \frac{yx}{r}\right)\frac{dV}{dr} = 0.$$

Thus the angular momentum does not change with time; it is conserved.

Conservation of electric charge is related to symmetry of the laws of nature under a group of global phase transformations. The description of this symmetry is less straightforward than for the others, and we forgo its description and the derivation of the conservation from it.

6.5.   COSMIC SPECULATIONS

Let us now return to the analysis of nature into the laws of nature and initial conditions. This division, whereby the laws of nature and initial conditions are considered as being independent, seems to be satisfactory for our understanding of nature from the microscopic scale to the macroscopic. It seems to work even for astronomical phenomena, where, for example, we think of different planetary systems or even galaxies as evolving according to the same laws of nature but from different initial conditions. This dichotomy fades away, however, at nature's extreme scales, at the subnuclear scale, the scale of elementary particles, and at the scale of the universe. Current scientific consensus holds that the universe had its beginning in a "big bang" in the far past. Can the evolution of the universe be analyzed in terms of initial conditions and subsequent development according to the laws of nature? Perhaps, if one considers our universe to be only one of a variety of possible universes, any one of which could have come into being at the time of the "big bang." Then the universe to which we have access would be the result of whatever chance initial conditions initiated its evolution. And the other possible universes? Are they but theoretical possibilities, or did they in some sense all come into being but are inaccessible to us? Such speculations, however intriguing, can never be proved by the traditional scientific method, since we cannot experiment with different sets of initial conditions for the universe and follow the evolution of each. This way of looking at things seems to assign to the laws of nature a certain primacy over the universe itself.

Another point of view, which some people find philosophically more satisfactory, is that, since our universe is, by very definition, the totality of all we have access to, it is meaningless to consider other possible universes, and we must honor our universe with the distinction of being the only one possible according to the laws of nature. But what are the laws of nature, if not a description of the evolution of the universe? What this point of view holds is that the "big bang," the universe, initial conditions, and the laws of nature are all intimately intertwined, are all aspects of a single, self-consistent situation. Another aspect of this situation is the variety of elementary particles found in nature. While cosmologists are asking why the universe is as it is on the cosmic scale, other scientists are asking why it is as it is on the subnuclear scale, why nature presents us with the kinds of elementary particles that it does and no other kinds. Is this phenomenon explainable on the basis of the known laws of nature or must new laws of nature be found that relate more directly to the "big bang" and its immediately subsequent developments, when the present variety of elementary particles presumably came into existence? The latter possibility seems the more probable. This point of view might be summarized by the statement that things are as they are because they cannot be otherwise. See [M37].

How is symmetry connected with all this? Every aspect of that self-consistent entity, the universe, exhibits various symmetries and approximate symmetries, and we must try to understand how they, too, fit into the all-encompassing picture. The following tale is possibly a small part of that picture.

For every allowed state of any elementary particle having nonzero mass the spatially inverted (i.e., transformed by point inversion) state is also allowed by nature. And the photon, although it is massless, also has this property. Now the law of nature called the strong interaction, which affects certain of the massive particles (but not all), possesses space-inversion symmetry. And the law of nature called the electromagnetic interaction, which involves the photon and all electrically charged elementary particles, all of which are massive, is also inversion symmetric. Thus we have the situation where certain laws of nature and the entities whose behavior they govern possess the same symmetry, space-inversion symmetry, also called P symmetry, where P (for parity) denotes the point-inversion transformation. In addition, the transformation of particle–antiparticle conjugation, denoted C, is a symmetry both of the strong and electromagnetic interactions and of the massive elementary particles and the photon; for every kind of such particle there is a kind that is the antiparticle of it (with certain ones, including the photon, being their own antiparticles), and all states allowed for particles are also allowed for their antiparticles and vice versa. So again we have the same symmetry, C symmetry, for certain laws of nature and the systems they apply to.

On the other hand, the neutrinos, massless like the photon, are deficient in these symmetries. The space-inversion image of an allowed neutrino state is not a state allowed by nature. It would be, if the image particle were an antineutrino, for the allowed states of antineutrinos are just the forbidden states of neutrinos. And the inverted states of antineutrinos are forbidden to antineutrinos but allowed for neutrinos. It is as if only left-handed neutrinos and right-handed antineutrinos are permitted (or, better, required) to exist by nature, while the existence of right-handed neutrinos and left-handed antineutrinos is excluded. Thus neither P nor C is a symmetry of neutrinos, while the combined transformation of point inversion and particle–antiparticle conjugation, denoted CP, is a symmetry transformation. Now the strong and electromagnetic interactions do not affect neutrinos, while the law of nature called the weak interaction does. (It also affects other particles.) The weak interaction possesses neither P symmetry nor C symmetry; it does possess CP symmetry.

Are the P and C symmetries of the objects upon which the strong and electromagnetic interactions act basically caused by the P and C symmetries of the interactions themselves, or vice versa, or are the interactions and the elementary particles, along with their symmetries, parts of a self-consistent situation with neither more basic than the other? And are the P and C asymmetries and CP symmetry of the neutrinos fundamentally a result of the P and C asymmetries and CP symmetry of the weak interaction, or vice versa, or is it all a self-consistent whole? And how does all this relate to conditions prevailing at the earliest stages of the evolution of the universe, when the kinds of elementary particles that we know today presumably came into being? Why were the other-handed versions of the neutrinos and antineutrinos not created then? Did the laws of nature even then preclude their existence? Or did the chance creation of the first neutrino–antineutrino pair as a left-handed neutrino and a right-handed antineutrino cause the as yet "plastic" laws of nature to "crystallize" into such a form that would from that moment onward allow proliferation of the existing versions of neutrino and antineutrino and forbid the existence of the other-handed versions (spontaneous symmetry breaking)? Or should we avoid the question by restructuring our concepts to make *CP* the only meaningful transformation, so that there would be no reason to consider the hypothetical possibility of other versions of neutrinos and antineutrinos? But then why are *P* and *C* separately relevant for the strong and electromagnetic interactions?

Our pursuit of symmetry and the understanding of symmetry carries us to the very frontiers of modern science and even to speculations out beyond.

Onward

This brings to a close our symmetry primer for scientists, and you should be well started on your way toward an understanding and appreciation of symmetry and its application in science. Keep in mind, though, that this was only an introduction; there is much more ahead. The fields where symmetry considerations have traditionally played an important role are quantum mechanics and crystallography. More recently symmetry considerations have grown in importance, sometimes considerably, in many additional fields, such as solid-state physics, nuclear physics, the physics of particles and fields (also called high-energy or elementary-particle physics), general relativity, physical chemistry, especially quantum chemistry, low-temperature physics, and thermodynamics. And even in less fundamental fields, such as biology and engineering, symmetry considerations are gaining in importance. I doubt whether there is a single field of scientific endeavor in which symmetry considerations cannot prove useful.

For more advanced study you are referred to the bibliography, and many additional references can be found in the books and articles listed there. For additional not necessarily very more advanced reading see [S24], [S17], and [S21]. The last reference includes a discussion of symmetry of composite systems.

Although our approach to symmetry in this book has been very "materialistic," in that we have studied the application of symmetry to science, i.e., to material systems and the laws governing them, you must not leave with the impression that this is all there is to symmetry. Symmetry has its "spiritual" sides also, especially symmetry in art. The wide applicability of symmetry is emphasized in [S17], and symmetry in art is discussed in [S24] and [S21]. See also [S19].

Bibliography

The principal purpose of this bibliography is to offer sources for parallel, supplementary, complementary, and subsequent reading. It consists of a compilation of books and articles that I have read, come across, read reviews of, seen references to, or written and consider suitable in content and level. No claim is made for completeness. Its makeup reflects my own view of the field. The entries are classified, in some cases rather arbitrarily, into four sections:

G—Abstract Group Theory. Little symmetry, if any; rare application; corresponds pretty well to Chapters 1 and 2.

S—General Symmetry. Usually little group theory, if any; no detailed application, such as that involving group representations; roughly corresponds to Chapter 3, with some relevance to Chapter 4.

A—Applied Symmetry and Group Theory. Sometimes with a lot of abstract group theory; most of the entries are presentations of the type for which this book is intended to supply conceptual foundations; thus this section mostly offers sequels to Chapter 4, but some of the entries are good for Chapter 4 itself.

M—Miscellaneous. Not classifiable under G, S, or A; material relevant to Chapters 5 and 6 is mostly found here.

References to entries have the form [Xi], where X = G, S, A, M and i = 1, 2, . . . , indicating the ith entry in section X.

For a broader and more extensive bibliography see my article "Resource letter SP-2: Symmetry and group theory in physics" [S18]. In spite of its title it is not limited strictly to physics and covers also chemistry, biology, and art.

G—Abstract Group Theory

G1   P. S. Alexandroff, *An Introduction to the Theory of Groups* (Blackie, Glasgow, 1959).

G2   B. Baumslag and B. Chandler, *Theory and Problems of Group Theory* (Schaum's Outline Series, McGraw-Hill, New York, 1968).

G3   A. W. Bell, *Algebraic Structures—Some Aspects of Group Structure* (George Allen & Unwin, London, 1966).

G4   G. Birkhoff and S. MacLane, *A Survey of Modern Algebra* (Macmillan, New York, 1965).

G5   F. J. Budden, *The Fascination of Groups* (Cambridge University Press, Cambridge, 1972).

G6   H. S. M. Coxeter and W. O. J. Moser, *Generators and Relations for Discrete Groups* (Springer-Verlag, Berlin, 1965).

G7   J. D. Dixon, *Problems in Group Theory* (Dover, New York, 1973).

G8   R. Gilmore, *Lie Groups, Lie Algebras, and Some of Their Applications* (Wiley, New York, 1974).

G9   I. Grossman and W. Magnus, *Groups and Their Graphs* (Random House, New York, 1964).

G10 C. A. Hollingsworth, *Vectors, Matrices, and Group Theory for Scientists and Engineers* (McGraw-Hill, New York, 1967).

G11 W. Ledermann, *Introduction to Group Theory* (Barnes & Noble, New York, 1973).

G12 E. S. Lyapin, A. Ya. Aizenshtat, and M. M. Lesokhin, *Exercises in Group Theory* (Plenum, New York, 1972).

G13 The Open University, *Groups I. Mathematics Foundation Course Unit 30* (Open University Press, Milton Keynes, U.K., 1971).

G14 The Open University, *Groups II. Mathematics Foundation Course Unit 33* (Open University Press, Milton Keynes, U.K., 1972).

G15 The Open University, *Group Axioms, Group Morphisms. Topics in Pure Mathematics Units 4 and 5* (Open University Press, Milton Keynes, U.K., 1973).

G16 The Open University, *Group Structure. Topics in Pure Mathematics Units 11 and 12* (Open University Press, Milton Keynes, 1973).

G17 E. M. Patterson and D. E. Rutherford, *Elementary Abstract Algebra* (Oliver & Boyd, Edinburgh, 1965).

G18 A. A. Sagle and R. E. Walde, *Introduction to Lie Groups and Lie Algebras* (Academic, New York, 1973).

G19 G. Stephenson, *Matrices, Sets and Groups* (American Elsevier, New York, 1966).

G20 D. A. R. Wallace, *Groups* (George Allen & Unwin, London, 1974).

G21 H. Weyl, *The Classical Groups, Their Invariants and Representations* (Princeton University Press, Princeton, 1946).

G22 T. A. Whitelaw, *An Introduction to Abstract Algebra* (Blackie, Glasgow, 1978).

S—General Symmetry

S1   I. Bernal, W. C. Hamilton, and J. S. Ricci, *Symmetry, A Stereoscopic Guide for Chemists* (Freeman, San Francisco, 1972).

S2   H. S. M. Coxeter, *Regular Polytopes* (Dover, New York, 1963).

S3   H. S. M. Coxeter, *Introduction to Geometry* (Wiley, New York, 1969), Chapters 2–5, 7, 10, 11, 15.

S4   L. Fejes Tóth, *Regular Figures* (Pergamon, Oxford, 1964).

S5   V. Fritsch, *Left and Right in Science and Life* (Humanities, Atlantic Highlands, NJ, 1968).

S6   M. Gardner, *Ambidextrous Universe: Mirror Asymmetry and Time Reversed Worlds* (Scribner, New York, 1979).

S7   L. Glasser, "Teaching symmetry, the use of decorations," *J. Chem. Educ.* **44,** 502–511 (1967).

S8   A. Holden, *Shapes, Space, and Symmetry* (Columbia University Press, New York, 1971).

S9   H. R. Jacobs, *Geometry* (Freeman, San Francisco, 1974), Chapter 5.

S10   M. A. Jaswon, *An Introduction to Mathematical Crystallography* (American Elsevier, New York, 1965).

S11   G. Karl, "On geometrical symmetry," *Am. J. Phys.* **35,** 98–101 (1967).

S12   H. Liebeck, *Algebra for Scientists and Engineers* (Wiley, New York, 1969).

S13   E. H. Lockwood and R. H. MacMillan, *Geometric Symmetry* (Cambridge University Press, Cambridge, 1978).

S14   A. L. Loeb, *Color and Symmetry* (Wiley, New York, 1971).

S15   J. Nicolle, *La Symétrie et Ses Applications* (Albin Michel, Paris, 1950); *La Symétrie* (Presses Universitaires de France, Paris, 1957).

S16   D. Park, "Resource letter SP-1 on symmetry in physics," *Am. J. Phys.* **36,** 577–584 (1968).

S17   J. Rosen, *Symmetry Discovered: Concepts and Applications in Nature and Science* (Cambridge University Press, Cambridge, 1975).

S18   J. Rosen, "Resource letter SP-2: Symmetry and group theory in physics," *Am. J. Phys.* **49,** 304–319 (1981).

S19   M. Senechal and G. Fleck (editors), *Patterns of Symmetry* (University of Massachusetts Press, Amherst, 1977).

S20   A. V. Shubnikov, N. V. Belov, et al., *Colored Symmetry* (Pergamon, Oxford, 1964).

S21   A. V. Shubnikov and V. A. Koptsik, *Symmetry in Science and Art* (Plenum, New York, 1974).

S22   A. F. Wells, *The Third Dimension in Chemistry* (Oxford University Press, Oxford, 1956).

S23   M. J. Wenninger, *Polyhedron Models for the Classroom* (National Council of Teachers of Mathematics (U.S.), Reston, VA, 1966); *Polyhedron Models* (Cambridge University Press, Cambridge, 1971).

S24   H. Weyl, *Symmetry* (Princeton University Press, Princeton, 1952).

S25   E. P. Wigner, "Violations of symmetry in physics," *Sci. Am.* **213** (6), 28–36 (December 1965).

S26   I. M. Yaglom, *Geometric Transformations,* Vols. I, II, III (Random House, New York, 1962, 1968, 1973).

A—Applied Symmetry and Group Theory

A1   V. Bargman, "Note on Wigner's theorem on symmetry operations," *J. Math. Phys.* **5,** 862–868 (1964).

A2   S. Bhagavantam, *Crystal Symmetry and Physical Properties* (Academic, New York, 1966).

A3   S. Bhagavantam and T. Venkatarayudi, *Theory of Groups and Its Application to Physical Problems* (Academic, New York, 1969).

A4   G. Birkhoff, *Hydrodynamics, A Study in Logic, Fact, and Similitude* (Dover, New York, 1950).

A5   G. D. Birkhoff, "The principle of sufficient reason," *Rice Inst. Pamphlet* **28**, 24–52 (1941), reprinted in *Collected Mathematical Papers* (Dover, New York, 1968), Vol. 3, pp. 778–804.

A6   R. R. Birss, *Symmetry and Magnetism* (North-Holland, Amsterdam, 1964).

A7   D. M. Bishop, *Group Theory and Chemistry* (Oxford University Press, Oxford, 1973).

A8   A. D. Boardman, D. E. O'Connor, and P. A. Young, *Symmetry and Its Applications in Science* (Wiley, New York, 1973).

A9   H. Boerner, *Representations of Groups* (North-Holland, Amsterdam, 1963).

A10 H. A. Buchdahl, *An Introduction to Hamiltonian Optics* (Cambridge University Press, Cambridge, 1970), Chapters 3–9.

A11 D. B. Chesnut, *Finite Groups and Quantum Theory* (Wiley, New York, 1974).

A12 C. D. H. Chisholm, *Group Theoretical Techniques in Quantum Chemistry* (Academic, New York, 1976).

A13 J. F. Cornwell, *Group Theory and Electronic Energy Bands in Solids* (North-Holland, Amsterdam, 1969).

A14 F. A. Cotton, *Chemical Applications of Group Theory* (Wiley, New York, 1963).

A15 A. P. Cracknell, *Applied Group Theory* (Pergamon, Oxford, 1968).

A16 A. P. Cracknell, *Group Theory in Solid State Physics* (Wiley, New York, 1975); also published as "Group theory in solid-state physics is not dead yet, *alias,* Some recent developments in the use of group theory in solid-state physics," *Adv. Phys.* **23**, 673–866 (1974).

A17 P. Curie, "Sur la symétrie dans les phénomènes physiques, symétrie d'un champ électrique et d'un champ magnétique," *J. Phys.* (3rd ser.) **3**, 393–415 (1894); reprinted in *Oeuvres de Pierre Curie* (Gauthier-Villars, Paris, 1908), pp. 118–141.

A18 G. Davidson, *Introductory Group Theory for Chemists* (Elsevier, London, 1971).

A19 J. D. Donaldson and S. D. Ross, *Symmetry and Stereochemistry* (International Textbook, London, 1972).

A20 P. B. Dorain, *Symmetry in Inorganic Chemistry* (Addison-Wesley, Reading, MA, 1965).

A21 J. McL. Emmerson, *Symmetry Principles in Particle Physics* (Oxford University Press, Oxford, 1972).

A22 L. M. Falikov, *Group Theory and Its Physical Applications* (University of Chicago Press, Chicago, 1966).

A23 J. R. Ferraro and J. S. Ziomek, *Introductory Group Theory and Its Application to Molecular Structure* (Plenum, New York, 1969).

A24 G. G. Hall, *Applied Group Theory* (American Elsevier, New York, 1967).

A25 L. H. Hall, *Group Theory and Symmetry in Chemistry* (McGraw-Hill, New York, 1969).

A26 M. Hamermesh, *Group Theory and Its Application to Physical Problems* (Addison-Wesley, Reading, MA, 1962).

A27   D. C. Harris and M. D. Bertolucci, *Symmetry and Spectroscopy, An Introduction to Vibrational and Electronic Spectroscopy* (Oxford University Press, New York, 1978).

A28   V. Heine, *Group Theory in Quantum Mechanics* (Pergamon, Oxford, 1960).

A29   R. M. Hochstrasser, *Molecular Aspects of Symmetry* (Benjamin, New York, 1966).

A30   F. M. Jaeger, *Lectures on the Principle of Symmetry and Its Applications in All Natural Sciences* (Elsevier, Amsterdam, 1917).

A31   H. H. Jaffé and M. Orchin, *Symmetry in Chemistry* (Wiley, New York, 1965).

A32   L. Jansen and M. Boon, *Theory of Finite Groups. Applications in Physics: Symmetry Groups of Quantum Mechanical Systems* (North-Holland, Amsterdam, 1967).

A33   T. Janssen, *Crystallographic Groups* (North-Holland, Amsterdam, 1973).

A34   D. F. Johnston, "Group theory in solid state physics," *Rep. Prog. Phys.* **23**, 66–153 (1960).

A35   A. W. Joshi, *Elements of Group Theory for Physicists* (Halsted, New York, 1974).

A36   J. Kepler, *The Six-Cornered Snowflake* (Oxford University Press, Oxford, 1966, originally published 1611).

A37   M. Lax, *Symmetry Principles in Solid State and Molecular Physics* (Wiley, New York, 1974).

A38   J. W. Leech and D. J. Newman, *How to Use Groups* (Halsted, New York, 1969).

A39   J. S. Lomont, *Applications of Finite Groups* (Academic, New York, 1959).

A40   G. Ya. Lyubarskii, *The Application of Group Theory in Physics* (Pergamon, Oxford, 1960).

A41   R. McWeeny, *Symmetry, An Introduction to Group Theory and Its Applications* (Pergamon, Oxford, 1963).

A42   P. H. E. Meijer and E. Bauer, *Group Theory: The Application to Quantum Mechanics* (North-Holland, Amsterdam, 1965).

A43   W. Miller, Jr., *Symmetry Groups and Their Applications* (Academic, New York, 1972).

A44   A. Nussbaum, *Applied Group Theory for Chemists, Physicists and Engineers* (Prentice-Hall, Englewood Cliffs, NJ, 1971).

A45   M. Orchin and H. H. Jaffé, *Symmetry, Orbitals, and Spectra (S.O.S.)* (Wiley, New York, 1971).

A46   L. F. Phillips, *Basic Quantum Chemistry* (Wiley, New York, 1965).

A47   I. V. Schensted, *A Course on the Application of Group Theory to Quantum Mechanics* (NEO, Ann Arbor, MI, 1965).

A48   D. S. Schonland, *Molecular Chemistry, An Introduction to Group Theory and Its Uses in Chemistry* (Van Nostrand-Reinhold, New York, 1965).

A49   R. Shaw, "Symmetry, uniqueness, and the Coulomb law of force," *Am. J. Phys.* **33**, 300–305 (1965).

A50   M. Tinkham, *Group Theory and Quantum Mechanics* (McGraw-Hill, New York, 1964).

A51   A. Vincent, *Molecular Symmetry and Group Theory, A Programmed Introduction to Chemical Applications* (Wiley, New York, 1977).

A52   B. L. van der Waerden, *Group Theory and Quantum Mechanics* (Springer-Verlag, Berlin, 1974).

A53   H. Weyl, *The Theory of Groups and Quantum Mechanics* (Dover, New York, 1931).

A54   E. P. Wigner, *Group Theory and Its Application to the Quantum Mechanics of Atomic Spectra* (Academic, New York, 1959).

A55   A. B. Wolbarst, *Symmetry and Quantum Systems* (Van Nostrand-Reinhold, New York, 1977).

A56   B. G. Wybourne, *Symmetry Principles in Atomic Spectroscopy* (Wiley, New York, 1970).

A57   B. G. Wybourne, *Classical Groups for Physicists* (Wiley, New York, 1974).

M—Miscellaneous

M1   J. Aharoni, *Lectures on Mechanics for Students of Physics and Engineering* (Oxford University Press, Oxford, 1972), Chapter 23.

M2   H. Alfvén, *Worlds–Antiworlds: Antimatter in Cosmology* (Freeman, San Francisco, 1966); "Antimatter and cosmology," *Sci. Am.* **216** (4), 106–114 (April 1967).

M3   J. M. Bailey, *Liberal Arts Physics: Invariance and Change* (Freeman, San Francisco, 1972), Section 2.

M4   H. Callen, "Thermodynamics as a science of symmetry," *Found. Phys.* **4,** 423–443 (1974).

M5   P. C. W. Davies, *The Physics of Time Asymmetry* (Surrey University Press, London, 1974).

M6   P. C. W. Davies, *Space and Time in the Modern Universe* (Cambridge University Press, Cambridge, 1977).

M7   P. A. M. Dirac, "The evolution of the physicist's picture of nature," *Sci. Am.* **208** (5), 45–53 (May 1963).

M8   G. H. Duffey, *Theoretical Physics: Classical and Modern Views* (Houghton Mifflin, Boston, 1973), Chapters 7–10.

M9   C. V. Durell, *Readable Relativity* (Harper & Row, New York, 1931).

M10   F. J. Dyson, "Mathematics in the physical sciences," *Sci. Am.* **211** (3), 128–146 (September 1964).

M11   E. B. Edwards, *Pattern and Design with Dynamic Symmetry* (Dover, New York, 1967, originally published 1932).

M12   G. Feinberg and M. Goldhaber, "The conservation laws of physics," *Sci. Am.* **209** (4), 36–45 (October 1963).

M13   R. P. Feynman, *The Character of Physical Law* (M.I.T. Press, Cambridge, 1965), esp. Chapters 3–5.

M14   R. P. Feynman, R. B. Leighton, and M. Sands, *The Feynman Lectures on Physics* (Addison-Wesley, Reading, MA, 1963), Vol. 1, Chapters 46, 52.

M15   G. Gamow, "The exclusion principle," *Sci. Am.* **201** (1), 74–86 (July 1959).

M16   G. Gamow, *One, Two, Three—Infinity: Facts and Speculations of Science* (Viking, New York, 1963).

M17   M. Gardner, "Can time go backward?," *Sci. Am.* **216** (1), 98–108 (January 1967).

M18   M. Gardner, *The Relativity Explosion* (Random House, New York, 1976).

M19   M. Ghyka, *The Geometry of Art and Life* (Sheed & Ward, New York, 1946).

M20   M. Gourdin, *Lagrangian Formalism and Symmetry Laws* (Gordon & Breach, New York, 1969).

M21   H. Haken, *Synergetics—An Introduction: Nonequilibrium Phase Transitions and Self-Organization in Physics, Chemistry and Biology* (Springer-Verlag, Berlin, 1977).

M22   J. Hambidge, *The Elements of Dynamic Symmetry* (Dover, New York, 1967, originally published 1926).

M23   P. Havas, "The connection between conservation laws and invariance groups: Folklore, fiction and fact," *Acta Phys. Austriaca* **38**, 145–167 (1973).

M24   R. M. F. Houtappel, H. Van Dam, and E. P. Wigner, "The conceptual basis and use of the geometric invariance principles," *Rev. Mod. Phys.* **37**, 595–632 (1965).

M25   G. Kepes (editor), *Module, Proportion, Symmetry, Rhythm* (Braziller, New York, 1966).

M26   L. D. Landau and G. W. Rumer, *What Is Relativity?* (Fawcett, New York, 1972).

M27   T. D. Lee, "Space inversion, time reversal and particle–antiparticle conjugation," *Phys. Today* **19** (3), 23–31 (March 1966).

M28   J. Leite Lopes, *Lectures on Symmetries* (Gordon & Breach, New York, 1969).

M29   C. H. MacGillavry, *Fantasy and Symmetry: The Periodic Drawings of M. C. Escher* (Abrams, New York, 1976).

M30   R. A. Mann, *The Classical Dynamics of Particles: Galilean and Lorentz Relativity* (Academic, New York, 1974), Chapters 3, 6.

M31   E. Merzbacher, *Quantum Mechanics* (Wiley, New York, 1970), Chaps. 16, 20.

M32   A. Messiah, *Quantum Mechanics* (North-Holland, Amsterdam, 1962), Vol. 2, Part 3.

M33   J. Neuberger, "Rotation matrix for an arbitrary axis," *Am. J. Phys.* **45**, 492–493 (1977).

M34   A. Palazzolo, "Formalism for the rotation matrix of rotations about an arbitrary axis," *Am. J. Phys.* **44**, 63–67 (1976).

M35   T. Poston and I. Stewart, *Catastrophe Theory and Its Applications* (Pitman, London, 1978).

M36   P. Renaud, "Sur une généralisation du principe de symétrie de Curie," *C. R. Acad. Sci. Paris* **200**, 531–534 (1935).

M37   J. Rosen, "Extended Mach principle," *Am. J. Phys.* **49**, 258–264 (1981).

M38   J. Rosen and Y. Freundlich, "Symmetry and conservation," *Am. J. Phys.* **46**, 1030–1041 (1978); and sequel, J. Rosen, "Symmetry and conservation: Inverse Noether's theorem and general formalism," *J. Phys. A* **13**, 803–813 (1980).

M39   E. Samuel, *Order: In Life* (Prentice-Hall, Englewood Cliffs, NJ, 1972).

M40   A. Sellerio, "Entropia, probabilità, simmetria," *Nuovo Cimento* **6**, 236–242 (1929); "Le simmetrie nella fisica," *Scientia* **58**, 69–80 (1935).

M41   P. S. Stevens, *Patterns in Nature* (Little, Brown, Boston, 1974).

M42   E. F. Taylor and J. A. Wheeler, *Spacetime Physics* (Freeman, San Francisco, 1966).

M43   R. Thom, *Structural Stability and Morphogenesis: An Outline of a General Theory of Models* (Benjamin, Reading, MA, 1975).

M44   D. W. Thompson, *On Growth and Form* (Cambridge University Press, Cambridge, 1942).

M45   L. L. Whyte, "Tendency towards symmetry in fundamental physical structures,"
      Nature **163**, 762–763 (1949).

M46   E. P. Wigner, *Symmetries and Reflections* (M.I.T. Press, Cambridge, 1967).

M47   C. N. Yang, *Elementary Particles* (Princeton University Press, Princeton, 1962).

M48   E. C. Zeeman, "Catastrophe theory," *Sci. Am.* **234** (4), 65–83 (April 1976).

Index

||||||||||||||||||||||||||||